BrightRED Study Guide

CfE ADVANCED Higher

MATHEMATICS

Linda Moon, Philip Moon and Dr Michael Green

First published in 2017 by:
Bright Red Publishing Ltd
1 Torphichen Street
Edinburgh
EH3 8HX

A CIP record for this book is available from the British Library.

ISBN 978-1-906736-72-9

With thanks to:
PDQ Digital Media Solutions Ltd, Bungay (layout); Ivor Normand (copy-edit).
Cover design and series book design by Caleb Rutherford – e i d e t i c.

Acknowledgements
Every effort has been made to seek all copyright-holders. If any have been overlooked, then Bright Red Publishing will be delighted to make the necessary arrangements.

Permission has been sought from all relevant copyright holders and Bright Red Publishing are grateful for the use of the following:

Horatius (CC BY-SA 3.0)[1] (p 47) Images licensed by Ingram Image (pp 73, 79, 81 & 87); CfE Advanced Higher Mathematics exam questions © Scottish Qualifications Authority (n.b. solutions do not emanate from the SQA).

[1] (CC BY-SA 3.0) https://creativecommons.org/licenses/by-sa/3.0/deed.en

Printed and bound in the UK by Martins the Printers.

MIX
Paper from
responsible sources
FSC® C013254
FSC
www.fsc.org

CONTENTS

INTRODUCTION

INTRODUCING CFE ADVANCED HIGHER MATHEMATICS

THE STRUCTURE AND AIM OF THIS BOOK

There is no short-cut to passing any course at Advanced Higher level. To obtain a good pass requires consistent, regular revision over the duration of the course. The aim of this book is to bring together, in one volume, with the aid of examples, concise coverage of the course material. The book should be used in conjunction with your course notes and the knowledge gained from attending classes.

In addition to an index at the end of the book, there is a summary skills list which details, in the order of the syllabus given by the SQA, each assessment standard of the course – unit by unit. There is a box alongside, which you might use to monitor and check your progress, perhaps as follows:

an empty box	I have not learned this
• a dot	we have covered this in class
– a dash	I understand this topic, but more practice is required
+ a cross	I have revised this area and I am confident with it.

On the website, you will find further tables (shorter summative versions) which may also be used to track your progress.

Regularly throughout this book are other features:

Don't Forget boxes flag up vital pieces of information that you need to remember and important things that you must be able to do, plus some helpful hints.

Things to Do and Think About sections contain practice questions to test your understanding. The solutions to these tasks are available on the *Bright Red Digital Zone* (www.brightredbooks.net).

Online references and tests direct you to the *Bright Red Digital Zone*, an online source of examples and solutions covering the entire syllabus. The examples range in length and complexity, and include questions in the same style and format as you will meet in your exams. Solutions will be provided to all examples. This is a useful resource as you study throughout the session.

Websites and **video links** can be used to enhance your knowledge. In addition, www.hsn.uk.net has summary notes, and the SQA website (www.sqa.org.uk) has information on the examination, past papers and solutions which may be useful.

Do remember that success in this hierarchical subject depends on your knowing and using skills from lower levels, such as Higher. Trigonometric identities need to be remembered as well as all the algebraic and calculus techniques you met.

COURSE STRUCTURE

The Advanced Higher Mathematics course is divided into three units (MAC, AAC, GPS):

Methods in algebra and calculus	Applications in algebra and calculus	Geometry, proof and systems of equations

• Algebraic skills
  • Partial fractions
• Differentiation
• Integration
• Differential equations
	• Algebraic skills
	  • Binomial theorem
	  • Complex numbers
	• Sequences and series
	• Summation formulae
	• Proof by induction
	• Functions
	• Calculus problems in context
		• Systems of linear equations
		• Matrices
		• Vectors in three dimensions
		• Geometric operations on complex numbers
		• Number theory and proof

ASSESSMENT

At present, the Advanced Higher Mathematics course is assessed in two ways:

Each of the three units, above, is assessed within your school and requires a pass (60%).

You will also take an externally assessed written examination consisting of a paper lasting three hours. The examination has an allocation of 100 marks. It is made up of shorter questions, usually twelve, that are generally worth between 3 and 7 marks. The final four questions will be longer and usually worth 40 marks in total.

The course is graded A (bands 1 and 2), B (bands 3 and 4), C (bands 5 and 6) or D (band 7) based on how well you do in the external examination.

During the next few years, the method for assessment will change. Up-to-date arrangements for assessment can be found at http://www.sqa.org.uk/sqa/48507.html

EXAMINATION HINTS

You do not need to answer the questions in order. It may be better to choose a question that you can answer easily, so that you settle your nerves. However, remember that the examiners will have attempted to put the shorter questions in order of difficulty, followed by the longer ones, also in order of difficulty. Thus question 12 might well be harder than question 16.

Remember: 100 marks in three hours, that is 1·8 minutes for every mark, or 18 minutes for a long question.

You will not be told in every question to 'show your working', but you need to remember to be accurate, to give detail and to illustrate your understanding in your working. However, you should simplify expressions and try to use concise and efficient methods where possible.

Rigour in mathematics is extremely important, as is communicating your conclusions and results with clarity.

REVISION TIPS

General advice

- **Don't leave your revision until the last minute**. When you are still learning new topics, revise the ones you have already covered.
- Study for periods of between 30 and 45 minutes, unless you are doing a complete paper.
- **Take short breaks**, away from your study area, to keep your level of concentration high.
- During your study leave, build treats and relaxation time into your revision timetable. This will help you to focus and help you stick to your plan.
- In the run-up to the exams, **eat well**, **exercise well** and **sleep well**.

Maths-specific revision tips

- The best way to revise mathematics is by doing it. There is a time for learning the necessary formulae and rules, but there is no substitute for practice.
- Once you have learned a topic or skill, try questions. Start off with straightforward questions, then Unit level, and progress to examination style. You want to test your knowledge of a topic by trying that discrete area of the syllabus, but you should progress to trying a mixture of past-paper questions. It is important that you start to recognise what to use and when, which skill to apply and where.
- Use the space between **Don't Forgets** in the margin of this book to add your own revision reminders.
- Mathematics is a subject to be practised often. Try to get into the habit of regularly doing mathematics. If you complete one extra question every night in addition to your normal homework and study time at school, you will reap the rewards. You will be able to ask for help the next day when the problem is fresh in your mind, so that you can tackle another question the next evening, and so you will quickly build up your knowledge and confidence.
- Mathematics also demands perseverance. There will come a time when you need to tackle a number of questions or a whole examination paper in one sitting. Time management is essential.
- Mathematics is different from other subjects in so many ways, and the good thing about revising it is that you can be active.

The best way to revise maths is to actively do it.

FORMULAE

On the next few pages are formulae that you should know. You should be very familiar with these so that you recognise them in less familiar contexts.

DIFFERENTIATION

Product rule:

$$\frac{d}{dx}(f(x)g(x)) = f'(x)g(x) + f(x)g'(x) \text{ or } \frac{d}{dx}(uv) = u'v + uv'$$

Quotient rule:

$$\frac{d}{dx}\left(\frac{f(x)}{g(x)}\right) = \frac{f'(x)g(x) - f(x)g'(x)}{(g(x))^2} \text{ or } \frac{d}{dx}\left(\frac{u}{v}\right) = \frac{u'v + uv'}{v^2}$$

$$\frac{dy}{dx} = \frac{1}{\frac{dx}{dy}}$$

Parametric equations

Given x and y are functions of t,

Gradient $\dfrac{dy}{dx} = \dfrac{\frac{dy}{dt}}{\frac{dx}{dt}}$

Speed $= \sqrt{\left(\dfrac{dx}{dt}\right)^2 + \left(\dfrac{dy}{dt}\right)^2}$

$$\frac{d^2y}{dx^2} = \frac{\frac{d^2y}{dt^2} \times \frac{dx}{dt} - \frac{dy}{dt} \times \frac{d^2x}{dt^2}}{\left(\frac{dx}{dt}\right)^3}$$

Standard derivatives	
$f(x)$	$f'(x)$
$\sin^{-1} x$	$\dfrac{1}{\sqrt{1-x^2}}$
$\cos^{-1} x$	$-\dfrac{1}{\sqrt{1-x^2}}$
$\tan^{-1} x$	$\dfrac{1}{1+x^2}$
$\tan x$	$\sec^2 x$
$\cot x$	$-\text{cosec}^2 x$
$\sec x$	$\sec x \tan x$
$\text{cosec } x$	$-\text{cosec } x \cot x$
$\ln x$	$\dfrac{1}{x}$
e^x	e^x

INTEGRATION

Standard integrals	
$f(x)$	$\int f(x)\, dx$
$\sec^2(ax)$	$\dfrac{1}{a}\tan(ax) + c$
$\dfrac{1}{\sqrt{a^2 - x^2}}$	$\sin^{-1}\left(\dfrac{x}{a}\right) + c$
$\dfrac{1}{a^2 + x^2}$	$\dfrac{1}{a}\tan^{-1}\left(\dfrac{x}{a}\right) + c$
$\dfrac{1}{x}$	$\ln|x| + c$
e^{ax}	$\dfrac{1}{a}e^{ax} + c$
$\tan x$	$\ln|\sec x| + c$
$\text{cosec } x$	$-\ln|\text{cosec } x + \cot x| + c$
$\cot x$	$\ln|\sin x| + c$
$\sec x$	$\ln|\sec x + \tan x| + c$

Integration by parts

$$\int f(x)\, g'(x)dx = f(x)g(x) - \int f'(x)\, g(x)dx \text{ or } \int uv'\, dx = uv - \int u'v\, dx$$

Volume of solid of revolution

About x-axis between $x = a$ and $x = b$: $V = \pi\displaystyle\int_a^b (f(x))^2\, dx$

About y-axis between $y = a$ and $y = b$: $V = \pi\displaystyle\int_a^b (f(y))^2\, dy$

DIFFERENTIAL EQUATIONS

First-order

For $\frac{dy}{dx} + P(x)y = Q(x)$, the integrating factor $I(x)$ is $e^{\int P(x)dx}$ and the solution is given by

$I(x)y = \int I(x)Q(x)dx.$

Second-order

Homogeneous equations: the complementary function

Nature of roots	Form of the general solution
Real unequal roots α and β	$y = Ae^{\alpha x} + Be^{\beta x}$
Real equal roots α and α	$y = Ae^{\alpha x} + Bxe^{\alpha x}$
Complex roots $p \pm qi$	$y = e^{px}(A\cos qx + B\sin qx)$

Non-homogeneous equations:

Right-hand side includes	Try for the particular integral
Polynomials: $y = ax + b$ $y = ax^2 + bx + c$ $y = ax^3 + bx^2 + c + d$	$y = Cx + D$ $y = Cx^2 + Dx + E$ $y = Cx^3 + Dx^2 + Ex + F$
$\sin ax$ or $\cos ax$	$y = C\sin ax + D\cos ax$
e^{ax}	$y = Ce^{ax}$

BINOMIAL THEOREM

$(a + b)^n = \sum_{r=0}^{n} \binom{n}{r}a^{n-r}b^r$ where $\binom{n}{r} = {}^nC_r = \frac{n!}{r!(n-r)!}$

Try to remember that this means that the general term for the binomial expansion

$(x + y)^n$ is ${}^nC_r x^{n-r} y^r$ which you need for finding a particular term.

COMPLEX NUMBERS

For the complex number $z = a + bi$,

- the **modulus** is given by $|z| = \sqrt{a^2 + b^2}$

- and the **argument** is given by $\tan\theta = \frac{b}{a}$ $-\pi < \theta < \pi$.

- The conjugate is $\bar{z} = a - bi$

De Moivre's theorem

$$[r(\cos\theta + i\sin\theta)]^n = r^n(\cos n\theta + i\sin n\theta)$$

SUMMATIONS

$\sum_{1}^{n} a = na$

$\sum_{r=1}^{n} r = \frac{n(n+1)}{2}$ $\sum_{r=1}^{n} r^2 = \frac{n(n+1)(2n+1)}{6}$ $\sum_{r=1}^{n} r^3 = \frac{n^2(n+1)^2}{4}$

FORMULAE

SEQUENCES AND SERIES

Arithmetic and geometric

	Arithmetic series	Geometric series
nth term	$u_n = a + (n-1)d$	$u_n = ar^{n-1}$
Sum of n terms	$S_n = \frac{1}{2}n\,[2a + (n-1)d]$	$S_n = \frac{a(1-r^n)}{1-r} \qquad r \neq 1$
Sum to infinity		$S_n = \frac{a}{1-r} \qquad |r| < 1$

Maclaurin expansion

$$f(x) = f(0) + f'(0)x + \frac{f''(0)x^2}{2!} + \frac{f'''(0)x^3}{3!} + \frac{f^{iv}(0)x^4}{4!} + \ldots$$

Useful to memorise:

$$e^x = 1 + x + \frac{x^2}{2!} + \frac{x^3}{3!} + \ldots + \frac{x^n}{n!} + \ldots$$

$$\sin x = x - \frac{x^3}{3!} + \frac{x^5}{5!} - \frac{x^7}{7!} + \ldots$$

$$\cos x = 1 - \frac{x^2}{2!} + \frac{x^4}{4!} - \frac{x^6}{6!} + \ldots$$

$$\tan^{-1} x = x - \frac{x^3}{3} + \frac{x^5}{5} - \frac{x^7}{7} + \ldots$$

$$\ln(1 + x) = x - \frac{x^2}{2} + \frac{x^3}{3} - \frac{x^4}{4} + \ldots$$

FUNCTIONS

Odd function: $f(-x) = -f(x)$	Even function: $f(-x) = f(x)$
(180° rotational symmetry)	(line symmetry about the y-axis)

MATRICES

		Determinant and inverse
2×2 matrices	$A = \begin{pmatrix} a & b \\ c & d \end{pmatrix}$	$\det A = ad - bc$ and $A^{-1} = \frac{1}{ad-bc}\begin{pmatrix} d & -b \\ -c & a \end{pmatrix}$
3×3 matrices	$A = \begin{pmatrix} a & b & c \\ d & e & f \\ g & h & i \end{pmatrix}$	$\det A = a\begin{vmatrix} e & f \\ h & i \end{vmatrix} - b\begin{vmatrix} d & f \\ g & i \end{vmatrix} + c\begin{vmatrix} d & e \\ g & h \end{vmatrix}$

$$(AB)^{-1} = B^{-1}A^{-1} \qquad (AB)' = B'A' \qquad \det AB = \det A \det B$$

Transformation matrices

Anti-clockwise rotation about origin through an angle θ	$\begin{bmatrix} \cos\theta & -\sin\theta \\ \sin\theta & \cos\theta \end{bmatrix}$	Clockwise rotation about origin through an angle θ	$\begin{bmatrix} \cos\theta & \sin\theta \\ -\sin\theta & \cos\theta \end{bmatrix}$
Reflection in the x-axis	$\begin{bmatrix} 1 & 0 \\ 0 & -1 \end{bmatrix}$	Reflection in the y-axis	$\begin{bmatrix} -1 & 0 \\ 0 & 1 \end{bmatrix}$
Reflection in the line $y = x$	$\begin{bmatrix} 0 & 1 \\ 1 & 0 \end{bmatrix}$	Reflection in the line $y = -x$	$\begin{bmatrix} 0 & -1 \\ -1 & 0 \end{bmatrix}$
Reflection in the origin	$\begin{bmatrix} -1 & 0 \\ 0 & -1 \end{bmatrix}$	Stretching/enlargement/ dilation ($k > 1$) or reduction ($k < 1$)	$\begin{bmatrix} k & 0 \\ 0 & k \end{bmatrix}$

VECTORS, LINES AND PLANES

Angle between two vectors: (Higher) $\mathbf{a.b} = |\mathbf{a}||\mathbf{b}|\cos\theta$

Equations of a line

Equation of a line through $\mathbf{a}$ (a, b, c) and with direction vector $\mathbf{d} = \begin{pmatrix} d_1 \\ d_2 \\ d_3 \end{pmatrix}$

Vector: $\mathbf{r} = \mathbf{a} + t\mathbf{d}$

Parametric: $x = a + td_1$ Symmetric: $\dfrac{x - a}{d_1} = \dfrac{y - b}{d_2} = \dfrac{z - c}{d_3} = \lambda$
$y = b + td_2$
$z = c + td_3$

Equations of a plane

Cartesian (non-parametric form): $(\mathbf{r} - \mathbf{a}).\mathbf{n} = 0$ or $\mathbf{r.n} = \mathbf{a.n}$ and if

$\mathbf{n} = \begin{pmatrix} l \\ m \\ n \end{pmatrix}$ and $\mathbf{a.n} = k$ then $lx + my + nz = k$

Vector: $\mathbf{r} = \mathbf{a} + s\mathbf{b} + t\mathbf{c}$

($\mathbf{r}$ and $\mathbf{a}$ are the respective position vectors of a general point and a given point on the plane. $\mathbf{b}$ and $\mathbf{c}$ are non-parallel vectors lying in the plane).

Parametric: $x = a_1 + sb_1 + tc_1$ $y = a_2 + sb_2 + tc_2$ $z = a_3 + sb_3 + tc_3$

Angle between two lines = acute angle between their direction vectors
Angle between two planes = acute angle between their normals
Angle between line and plane = 90° − (acute angle between $\mathbf{n}$ and $\mathbf{d}$)

Cross (vector) product:

$$\mathbf{a} \times \mathbf{b} = |\mathbf{a}||\mathbf{b}|\sin\theta\,\hat{\mathbf{n}} = \begin{vmatrix} \mathbf{i} & \mathbf{j} & \mathbf{k} \\ a_1 & a_2 & a_3 \\ b_1 & b_2 & b_3 \end{vmatrix} = \mathbf{i}\begin{vmatrix} a_2 & a_3 \\ b_2 & b_3 \end{vmatrix} - \mathbf{j}\begin{vmatrix} a_1 & a_3 \\ b_1 & b_3 \end{vmatrix} + \mathbf{k}\begin{vmatrix} a_1 & a_2 \\ b_1 & b_2 \end{vmatrix}$$

where $\hat{\mathbf{n}}$ is the normal to the plane containing $\mathbf{a}$ and $\mathbf{b}$.

Scalar triple product:

$$\mathbf{a}.(\mathbf{b} \times \mathbf{c}) = \begin{vmatrix} a_1 & a_2 & a_3 \\ b_1 & b_2 & b_3 \\ c_1 & c_2 & c_3 \end{vmatrix}$$

Finally, trigonometric identities and exact values that you should know from Higher:

	Essential formulae to know **off by heart** for the exam (G)	Others that may be useful for homework/classwork etc.
Links between ratios	$\cos^2 A + \sin^2 A = 1$ $\tan A = \dfrac{\sin A}{\cos A}$	$1 + \tan^2 A = \sec^2 A$ $\cot^2 A + 1 = \operatorname{cosec}^2 A$
Squared	$\cos^2 x = \frac{1}{2}(1 + \cos 2x)$ $\sin^2 x = \frac{1}{2}(1 - \cos 2x)$	
Compound angle	$\sin(A \pm B) = \sin A \cos B \pm \cos A \sin B$ $\cos(A \pm B) = \cos A \cos B \mp \sin A \sin B$	$\tan(A \pm B) = \dfrac{\tan A \pm \tan B}{1 \mp \tan B \tan A}$
Double angle	$\sin(2A) = 2\sin A \cos A$ $\cos(2A) = \cos^2 A - \sin^2 A$	$\tan(2A) = \dfrac{2\tan A}{1 - \tan^2 A}$

	0	$\frac{\pi}{6}$	$\frac{\pi}{4}$	$\frac{\pi}{3}$	$\frac{\pi}{2}$	π	$\frac{3\pi}{2}$	2π
sin	0	$\frac{1}{2}$	$\frac{1}{\sqrt{2}}$	$\frac{\sqrt{3}}{2}$	1	0	-1	0
cos	1	$\frac{\sqrt{3}}{2}$	$\frac{1}{\sqrt{2}}$	$\frac{1}{2}$	0	-1	0	1
tan	0	$\frac{1}{\sqrt{3}}$	1	$\sqrt{3}$	undef.	0	undef.	0

Negative facts:

$\sin(-\theta) = -\sin(\theta)$ $\cos(-\theta) = \cos(\theta)$ $\tan(-\theta) = -\tan(\theta)$

ALGEBRA

PARTIAL FRACTIONS MAC

Certain types of rational functions $\frac{p(x)}{q(x)}$, where p and q are polynomials in x, can be decomposed into partial fractions. This can be useful for integrating or differentiating this type of function. For examination purposes, q can be:

- quadratics or cubics which can easily be factorised into linear factors, or
- cubics which can be factorised into a product of a linear factor and an irreducible quadratic factor.

The quadratic/cubic may have a repeated linear factor.

LINEAR FACTORS

Example: 1.1

Express $\frac{x-5}{x^2-x-2}$ in partial fractions.

Solution:

$\frac{x-5}{x^2-x-2} = \frac{x-5}{(x+1)(x-2)} = \frac{A}{x+1} + \frac{B}{x-2}$.

Multiplying through by $(x+1)(x-2)$ gives

$x - 5 = A(x-2) + B(x+1)$.

$x = 2$ gives $B = -1$, $x = -1$ gives $A = 2$, so:

$\frac{x-5}{x^2-x-2} = \frac{2}{x+1} - \frac{1}{x-2}$.

LONG DIVISION

To obtain the partial fraction decomposition of $\frac{p(x)}{q(x)}$ when the degree of p is $\geqslant$ the degree of q, use long division to obtain a remainder of the form $\frac{r(x)}{q(x)}$, where the degree of r is $<$ the degree of q.

Partial fractions can now be performed on $\frac{r(x)}{q(x)}$.

Using long division

Example: 1.2

Express $\frac{2x^2-x-9}{x^2-x-2}$ in partial fractions.

Solution:

$\frac{2x^2-x-9}{x^2-x-2} = 2 + \frac{x-5}{x^2-x-2}$ —— Long division

$$\begin{array}{r} 2 \\ x^2-x-2 \overline{)\, 2x^2-x-9} \\ 2x^2-2x-4 \\ \hline x-5 \end{array}$$

Because the highest power of x is now less than 2, the process stops, giving the result shown.

Alternatively, this could be done by inspection.

Now proceeding as in example 1.1, we get:

$\frac{2x^2-x-9}{x^2-x-2} = 2 + \frac{2}{x+1} - \frac{1}{x-2}$.

 DON'T FORGET

Long division is required for $\frac{p(x)}{q(x)}$ when deg $p \geq$ deg q.

REPEATED FACTOR

Example: 1.3

Express $\frac{3x^2 - 11x + 4}{(x + 1)(x - 2)^2}$ in partial fractions.

Solution:

$$\frac{3x^2 - 11x + 4}{(x + 1)(x - 2)^2} = \frac{A}{(x + 1)} + \frac{B}{(x - 2)} + \frac{C}{(x - 2)^2}$$

$$3x^2 - 11x + 4 = A(x - 2)^2 + B(x + 1)(x - 2) + C(x + 1)$$

$x = -1$ gives $A = 2$, $x = 2$ gives $C = -2$.

Comparing the coefficient of x^2 on both sides gives $3 = A + B$, from which $B = 1$.

Therefore, $\frac{3x^2 - 11x + 4}{(x + 1)(x - 2)^2} = \frac{2}{x + 1} + \frac{1}{x - 2} - \frac{2}{(x - 2)^2}$

DON'T FORGET

If $q(x)$ contains a factor $(x - a)^2$, you need $\frac{A}{x - a} + \frac{B}{(x - a)^2}$

IRREDUCIBLE FACTOR

If $q(x)$ contains an irreducible factor such as $x^2 + a^2$, you need $\frac{Ax + B}{x^2 + a^2}$.

For $x^2 + x + 1$, you need $\frac{Ax + B}{x^2 + x + 1}$.

DON'T FORGET

Use $Ax + B$ for irreducible quadratics.

Example: 1.4

Express $\frac{2x^2 + x + 27}{x^3 + 9x}$ in partial fractions.

Solution:

$$\frac{2x^2 + x + 27}{x(x^2 + 9)} = \frac{A}{x} + \frac{Bx + C}{x^2 + 9}$$

$$\therefore 2x^2 + x + 27 = A(x^2 + 9) + x(Bx + C)$$

Comparing coefficients:

x^2: $2 = A + B$ $\therefore A = 3, B = -1, C = 1$

x: $1 = C$

x^0: $27 = 9A$

$$\frac{2x^2 + x + 27}{x^3 + 9x} = \frac{3}{x} + \frac{-x + 1}{x^2 + 9} \left[\text{or } \frac{3}{x} + \frac{1 - x}{x^2 + 9} \right]$$

ONLINE

Learn more about partial fractions by following the link at www.brightredbooks.net

THINGS TO DO AND THINK ABOUT

1 Express $\frac{x - 2}{3x^2 + 10x + 3}$ in partial fractions.   2

2 Express $\frac{1}{x^3 + 4x}$ in partial fractions.   4

3 Express $\frac{x^3 + x^2 + 2}{(x + 1)^2}$ in partial fractions.   5

4 Express $\frac{3x^2 - 13x + 50}{(x - 2)(x - 3)}$ in the form $A + \frac{B}{x - 2} + \frac{C}{x - 3}$ where $A, B, C \in \mathbb{Z}$.
State the values of A, B and C.   4

5 Describe the behaviour of the graph of function $f(x) = \frac{2x^3 + 3x^2 - 14x + k}{x^2 + 4x + 3}$,
where $k \in \mathbb{N}$ for large, positive values of x.   5

6 Express $\frac{x^3 + 2}{x(x^2 + 1)}$ in partial fractions.   5

VIDEO LINK

Check out the clip at www.brightredbooks.net for more on this topic.

ONLINE TEST

Test yourself on partial fractions at www.brightredbooks.net

FACTORIALS AND BINOMIAL COEFFICIENTS
AAC

The factorial operation is encountered in many areas of mathematics, notably in combinatorics, algebra and mathematical analysis. Its most basic occurrence is related to the fact that there are $n!$ ways to arrange n distinct objects into a sequence. This fact was known to Indian scholars at least as early as the 12th century.

The notation $n!$ was introduced by Christian Kramp in 1808.

DON'T FORGET

$n! = n \times (n - 1)!$

DON'T FORGET

0! is defined to be 1.

FACTORIALS

When n is a positive integer, the **factorial** of n is denoted by $n!$, where:

$$n! = n \times (n - 1) \times (n - 2) \times ... \times 2 \times 1$$

Your calculator should have an $n!$ button – but, because the factorial increases dramatically, it probably won't be able to calculate factorials above $n = 69$.

Example: 1.5

Evaluate 4!

Solution:

$4! = 4 \times 3 \times 2 \times 1 = 24$

Example: 1.6

Evaluate $\frac{12!}{10!}$

Solution:

$\frac{12!}{10!} = \frac{12 \times 11 \times 10!}{10!} = 12 \times 11 = 132$

Example: 1.7

Show that $(n - 1)! = (n - 1)(n - 2)!$

Solution:

$$(n - 1)! = (n - 1) \times (n - 2) \times ... \times 1$$
$$(n - 1)(n - 2)!$$

Example: 1.8

Show that $\frac{(n + 1)!}{(n - 1)!} = (n + 1)n$.

Solution:

$$\frac{(n + 1)!}{(n - 1)!} = \frac{(n + 1)(n)(n - 1)!}{(n - 1)!} = (n + 1)n$$

THE BINOMIAL COEFFICIENT $\binom{n}{r}$

When n is a positive integer, and r is an integer with $0 \leqslant r \leqslant n$, $\binom{n}{r} = \frac{n!}{r!(n-r)!}$ where $\binom{n}{r}$ is called the **binomial coefficient**.

Example: 1.9

Evaluate $\binom{5}{2}$.

Solution:

$$\binom{5}{2} = \frac{5!}{2!3!} = \frac{5 \times 4 \times 3 \times 2 \times 1}{2 \times 1 \times 3 \times 2 \times 1} = \frac{5 \times 4 \times 3!}{2 \times 3!} = 10$$

An alternative notation for the binomial coefficient $\binom{n}{r}$ is ^nC_r or $_nC_r$.

Check your calculator to see which notation is used.

If your calculator doesn't have factorial or binomial coefficient buttons (or if it packs up during the examination), there will always be factors which cancel to give a whole number.

Example: 1.10

Evaluate $\binom{25}{3}$.

Solution:

$$\binom{25}{3} = \frac{25!}{3!22!} = \frac{25 \times 24 \times 23 \times 22 \times \ldots}{3 \times 2 \times 1 \times 22 \times \ldots} = \frac{25 \times \overset{4}{\cancel{24}} \times 23 \times \cancel{22!}}{\underset{1}{6} \times \cancel{22!}} = 4 \times 25 \times 23 = 2300$$

Example: 1.11

Solve the equation $\binom{n}{3} = 2\binom{n}{2}$.

Solution:

Method 1

$$\frac{n!}{3!(n-3)!} = \frac{2 \times n!}{2!(n-2)!}$$

so, $(n-2)! = 3!(n-3)!$

$(n-2)(n-3)! = 6(n-3)!$

$n - 2 = 6$

$n = 8$

Method 2

$\binom{n}{3} = \frac{1}{6}n(n-1)(n-2)$ and $\binom{n}{2} = \frac{n(n-1)}{2}$

so, $\frac{1}{6}n(n-1)(n-2) = n(n-1)$

Dividing by $n(n-1)$, which is non-zero because $n \geq 3$ for $\binom{n}{3}$ to be defined, gives:

$\frac{n-2}{6} = 1 \Rightarrow n = 8$.

THINGS TO DO AND THINK ABOUT

1 Show that $\binom{n+1}{4} - \binom{n}{4} = \binom{n}{3}$. 4

2 Solve the equation $8\binom{n}{2} = 3\binom{n}{3}$. 3

3 Prove that $\binom{n+1}{p+1} - \binom{n}{p+1} = \binom{n}{p}$. 5

4 Solve for n:

$$\binom{n}{2} + \binom{n+1}{n} = 2\binom{n-1}{2} - 1$$

where $n \in \mathbb{N}$: $n \geqslant 3$. 4

DON'T FORGET

$\binom{n}{r} = \binom{n}{n-r}$

DON'T FORGET

$\binom{n}{0} = \binom{n}{n} = 1$ for all n

ONLINE

Follow the links at www.brightredbooks.net for more on factorials and binomial coefficients.

VIDEO LINK

Watch the video clip at www.brightredbooks.net to learn more.

DON'T FORGET

$\binom{n}{2} = \frac{1}{2}n(n-1)$.

ONLINE TEST

Test yourself on factorials and binomial coefficients at www.brightredbooks.net

THE BINOMIAL THEOREM ⒶⒶⒸ

The binomial theorem states that:

$$(a + b)^n = a^n + \binom{n}{1}a^{n-1}b + \binom{n}{2}a^{n-2}b^2 + \dots + \binom{n}{n-1}ab^{n-1} + b^n$$

for any positive integer n. The binomial theorem is used to describe the expansion of powers of a binomial, using a sum of terms. Coefficients in the expansion are binomial coefficients.

Example: 1.12

Obtain the binomial expansion of $(x + 2)^5$.

Solution:

$$(x + 2)^5 = x^5 + \binom{5}{1}x^4 2 + \binom{5}{2}x^3 2^2 + \binom{5}{3}x^2 2^3 + \binom{5}{4}x 2^4 + 2^5$$

$$(x + 2)^5 = x^5 + 10x^4 + 40x^3 + 80x^2 + 80x + 32.$$

Example: 1.13

Obtain the binomial expansion of $(2a - 3)^4$.

Solution:

$$(2a - 3)^4 = {}^4C_0(2a)^4(-3)^0 + {}^4C_1(2a)^3(-3)^1 + {}^4C_2(2a)^2(-3)^2 + {}^4C_3(2a)^1(-3)^3 + {}^4C_4(2a)^0(-3)^4$$

$$= (2a)^4 + 4(2a)^3(-3) + 6(2a)^2(-3)^2 + 4(2a)(-3)^3 + (-3)^4$$

$$= 16a^4 - 96a^3 + 216a^2 - 216a + 81.$$

Note that, when you are asked to write out the full expansion in assessments, the expansion indices will be no higher than 5.

Example: 1.14

Obtain the term independent of x in the expansion of $\left(x + \frac{3}{x}\right)^4$.

Solution:

The general term is:
$$\binom{4}{r}x^{4-r}\left(\frac{3}{x}\right)^r = \binom{4}{r}3^r x^{4-2r}.$$
We want the index of x to be 0, so:
$$4 - 2r = 0 \Rightarrow r = 2.$$
Thus the constant term is:
$$\binom{4}{2}3^2 = 6 \times 9 = 54.$$

Example: 1.15

Obtain the coefficient of x^2 in the expansion of $\left(3x^2 - \frac{1}{x^3}\right)^{11}$

Solution:

The general term is:
$$\binom{11}{r}(3x^2)^{11-r}\left(-\frac{1}{x^3}\right)^r = \binom{11}{r}3^{11-r}(-1)^r x^{22-5r}.$$
We need $22 - 5r = 2 \Rightarrow r = 4$.
Hence the coefficient of x^2 is:
$$\binom{11}{4}3^{11-4}(-1)^4 = \frac{11 \times 10 \times 9 \times 8 \times 3^7}{1 \times 2 \times 3 \times 4} = 721\,710.$$

PASCAL'S TRIANGLE

Pascal's triangle is a triangular array of binomial coefficients.

The first few rows of Pascal's triangle are shown here:

Pascal's triangle depends on the result $\binom{n+1}{r} = \binom{n}{r} + \binom{n}{r-1}$.

You should know how to prove this result.

Note that $\binom{4}{0} = 1$; $\binom{4}{1} = 4$; $\binom{4}{2} = 6$; $\binom{4}{3} = 4$; $\binom{4}{4} = 1$.

ONLINE

Learn more about Pascal's triangle by following the links at www.brightredbooks.net

DON'T FORGET

$\binom{n+1}{r} = \binom{n}{r} + \binom{n}{r-1}$

THINGS TO DO AND THINK ABOUT

1   Expand $(a^3 + 2)^4$ by the binomial theorem.   3

2   Write down and simplify the general term in the expansion of $\left(2x^3 + \frac{1}{x}\right)$.   3
    Hence, or otherwise, obtain the term in x^5.   2

3   Write down and fully simplify the general term in the binomial expansion of
    $\left(18x^2 - \frac{1}{12x}\right)^7$.
    Hence, obtain the term in x^5.   6

4   Write down the general term in the binomial expansion of $\left(3x - \frac{2}{x^2}\right)^6$.
    Hence obtain the term independent of x.   5

5   By writing $(1 + x)^{n+1} = (1 + x)^n (1 + x)$ and considering the coefficient of
    x^r $(1 \leqslant r \leqslant n)$ on both sides, prove that
    $\binom{n+1}{r} = \binom{n}{r} + \binom{n}{r-1}$ 4

6   Prove that
    $\binom{n}{r} + 2\binom{n}{r+1} + \binom{n}{r+2} = \binom{n+2}{r+2}$.   4

ONLINE TEST

Test yourself on the binomial theorem at www.brightredbooks.net

7   Prove that
    $\binom{n}{1} + \binom{n}{2} + \ldots + \binom{n}{n-1} = 2^n - 2$.   4

DIFFERENTIATION

STANDARD DERIVATIVES AND RULES FOR DIFFERENTIATION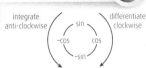

STANDARD DERIVATIVES

Although many of these derivatives now appear in the formulae sheet, you should know all of the derivatives in the following table or be able to derive them. For example: $\frac{d}{dx}(\tan x) = \frac{d}{dx}\left(\frac{\sin x}{\cos x}\right)$, and then use the quotient rule (or the product rule) as below.

Derivatives to learn

$f(x)$	$f'(x)$
e^x	e^x
$\ln x \ (or \log_e x)$	$\frac{1}{x}$
$\tan x$	$\sec^2 x$
$\sec x$	$\sec x \tan x$
$\operatorname{cosec} x$	$-\frac{\cos x}{\sin^2 x} = -\cot x \operatorname{cosec} x$
$\cot x$	$-\operatorname{cosec}^2 x$
$\sin^{-1} x, \sin^{-1} ax$	$\frac{1}{\sqrt{1-x^2}}, \frac{a}{\sqrt{1-(ax)^2}}$
$\cos^{-1} x, \cos^{-1} ax$	$\frac{-1}{\sqrt{1-x^2}}, \frac{-a}{\sqrt{1-(ax)^2}}$
$\tan^{-1} x, \tan^{-1} ax$	$\frac{1}{1+x^2}, \frac{a}{1-(ax)^2}$

As with $\sin x$, $\cos x$ from Higher, these new trig derivative results require x to be expressed in radians.

Similarly, at Higher you met straightforward applications of the chain rule:

$$\frac{d}{dx}(\sin 4x) = 4 \cos 4x$$

Example: 2.1

(a) $f(x) = \operatorname{cosec} 2x$

$f'(x) = -2\cot 2x \operatorname{cosec} 2x$

(b) $f(x) = \sec x^2$

$f'(x) = 2x \sec x^2 \tan x^2$

RULES FOR DIFFERENTIATION

Product rule

The product rule can be expressed as:

$(f(x)g(x))' = f'(x)g(x) + f(x)g'(x)$

and is used to find the derivatives of products of functions.

Example: 2.2

$\frac{d}{dx}(4x \sin x) = 4 \sin x + 4x \cos x$

contd

DON'T FORGET

integrate anti-clockwise / differentiate clockwise — sin, cos, −cos, −sin

DON'T FORGET

You should be able to recognise and use different notations:
- functional notation: $f(x)$, $f'(x)$, $f''(x)$ ...
- Leibniz notation: $\frac{dy}{dx}, \frac{d^2y}{dx^2}$ ...

DON'T FORGET

e^{3x} can be written as $\exp(3x)$.

DON'T FORGET

$\tan^{-1} x$ is **not** $\frac{1}{\tan x}$

DON'T FORGET

Remember
$f'(x) = \lim_{h \to 0} \frac{f(x+h) - f(x)}{h}$

DON'T FORGET

$(fg)' = f'g + fg'$

Quotient rule

The quotient rule can be expressed as:

$$\left(\frac{f(x)}{g(x)}\right)' = \frac{f'(x)\,g(x) - f(x)\,g'(x)}{(g(x))^2}$$

and is used to find the derivatives of quotients of functions.

Example: 2.3

$$\frac{d}{dx}\left(\frac{\sin x}{\ln x}\right) = \frac{\cos x \ln x - \sin x \cdot \frac{1}{x}}{(\ln x)^2}$$

$$= \frac{x \cos x \ln x - \sin x}{x(\ln x)^2}$$

Chain rule

The chain rule can be expressed as:

$$(f(g(x)))' = f'(g(x)) \cdot g'(x)$$

and is used for differentiating a composition of functions.

Example: 2.4

$$\frac{d}{dx}(e^{\tan x}) = e^{\tan x} \times \sec^2 x$$

Example: 2.5

$$\frac{d}{dx}(\tan^{-1}(1 - 2x)) = \frac{1}{1 + (1 - 2x)^2} \times (-2)$$

$$= \frac{-2}{2 - 4x + 4x^2}$$

$$= -\frac{1}{1 - 2x + 2x^2}$$

THINGS TO DO AND THINK ABOUT

1 Define $f(x) = \sin^2 x \cdot e^{-\tan x}$. Obtain $f'(x)$ and evaluate $f'\left(\frac{\pi}{4}\right)$.   3, 1

2 Given $f(x) = x^2 \tan 3x \left(0 < x < \frac{\pi}{6}\right)$, obtain $f'(x)$.   3

3 Given $y = \frac{e^x}{1 + 2x}\left(x \neq -\frac{1}{2}\right)$, obtain the value of x for which $\frac{dy}{dx} = 0$.   3

4 The position of a particle along the x-axis is modelled by:
$S(t) = t^4 - 2t^3 + \frac{3}{2}t^2, \ t \geqslant 0$.
t is the number of seconds elapsed. After how long will the particle first have constant velocity?   5

DON'T FORGET

$$\left(\frac{f}{g}\right)' = \frac{f'g + fg'}{g^2}$$

DON'T FORGET

When using the quotient rule, square the denominator $g(x)$.

DON'T FORGET

Tidy up solutions.

DON'T FORGET

$$\frac{dy}{dx} = \frac{1}{\frac{dx}{dy}}$$

ONLINE

Follow the links at www.brightredbooks.net to explore this topic further.

VIDEO LINK

Check out the clip at www.brightredbooks.net

ONLINE TEST

Test yourself on standard derivatives and the rules for differentiation at www.brightredbooks.net

FURTHER DIFFERENTIATION 1

HIGHER DERIVATIVES

If y is a function of x, so is $\frac{dy}{dx}$, and this can also be differentiated with respect to x.

We denote $\frac{d}{dx}\left(\frac{dy}{dx}\right)$ by $\frac{d^2y}{dx^2}$, and call this the second derivative of y with respect to x.

In the same way, we can form the third derivative, $\frac{d^3y}{dx^3}$, and so on.

If we are using the notation $f'(x)$, then the second derivative is denoted by $f''(x)$ and so on.

$$\frac{d^{n+1}y}{dx^{n+1}} = \frac{d}{dx}\left(\frac{d^ny}{dx^n}\right)$$

Example: 2.6

$y = (3x + 1)^4$

$\frac{dy}{dx} = 12(3x + 1)^3$

$\frac{d^2y}{dx^2} = 108(3x + 1)^2.$

Example: 2.7

If $\frac{d^2y}{dx^2} = \sin^4 2x$,

then $\frac{d^3y}{dx^3} = 8 \cos 2x \sin^3 2x.$

Example: 2.8

A car travels along a straight road starting from rest at time $t = 0$ at a point A.
After t seconds, it has travelled a distance $\frac{t^3}{3} + t^2$ metres.
At what time will the car break a speed limit of $25\,\mathrm{m\,s^{-1}}$?

Solution:

Let $x = \frac{t^3}{3} + t^2$ be the distance travelled, so that the speed $\frac{dx}{dt} = t^2 + 2t$. We need the positive value of t for which $t^2 + 2t = 25$. This gives $t = \sqrt{26} - 1 = 4.1\,\mathrm{s}$ to 1 decimal place.

The car will break the speed limit after $4.1\,\mathrm{s}$. (Maxima/minima questions appear in the Functions chapter, pp. 24–33.)

DERIVATIVES OF INVERSE FUNCTIONS

Let $y = f(x)$ be a function of x such that each value of y is given by a unique value of x. Then x can be regarded as a function of y: $x = f^{-1}(y)$. This is the **inverse function** of f. (You met inverse functions at Higher: see BrightRED Study Guide, pp. 20–21, 26).

To obtain $\frac{d}{dx}[f^{-1}(x)]$ for a given f, proceed as follows:

Let $y = f^{-1}(x)$ so that $x = f(y)$ and $\frac{dx}{dy} = f'(y) \Rightarrow \frac{dy}{dx} = \frac{1}{f'(y)}.$

Now express this in terms of x, simplifying where possible.

contd

DON'T FORGET

Give exact values in your answers rather than decimal approximations.

DON'T FORGET

Any of these differentiation skills could be tested in contextualised questions such as optimisation or displacement/velocity/acceleration.

DON'T FORGET

$f^{-1}(x)$ is not the same as $\frac{1}{f(x)}.$

Example: 2.9

Obtain the derivative of $y = \sin^{-1} x$ $-1 < x < 1$

Solution:

$x = \sin y$

$\frac{dx}{dy} = \cos y$

$\frac{dy}{dx} = \frac{1}{\cos y}$

Using $\cos y = \sqrt{1 - \sin^2 y}$ and $\sin y = x$

gives $\frac{dy}{dx} = \frac{1}{\sqrt{1 - x^2}}$

So, $\frac{d}{dx}(\sin^{-1} x) = \frac{1}{\sqrt{1 - x^2}}$

Example: 2.10

Obtain the derivative of $\sec^{-1} x$, where $x \geq 1$.

Solution:

$y = \sec^{-1} x \Rightarrow x = \sec y$, so $\frac{dx}{dy} = \sec y \tan y$. Using $\tan y = \sqrt{\sec^2 y - 1}$ gives:

$\frac{dx}{dy} = x\sqrt{x^2 - 1}$ so that $\frac{d}{dx}(\sec^{-1} x) = \frac{1}{x\sqrt{x^2 - 1}}$

Example: 2.11

Obtain the derivative of $y = \cos^{-1} 3x$ $-\frac{1}{3} < x < \frac{1}{3}$

Solution:

$x = \frac{1}{3}\cos y$

$\frac{dx}{dy} = -\frac{1}{3}\sin y$

$\frac{dy}{dx} = -\frac{3}{\sin y} = -\frac{3}{\sqrt{1 - \cos^2 y}}$

So, $\frac{dy}{dx} = -\frac{3}{\sqrt{1 - (3x)^2}} = -\frac{3}{\sqrt{1 - 9x^2}}$

In the examination, you can use the formulae that you have learned and which are given in the formulae sheet, as in the example below.

Example: 2.12

Differentiate $y = x^2 \sin^{-1} \frac{x}{5}$

Solution:

$\frac{dy}{dx} = 2x \sin^{-1} \frac{x}{5} + x^2 \times \frac{1}{5} \frac{1}{\sqrt{1 - (\frac{x}{5})^2}}$

$= 2x \sin^{-1} \frac{x}{5} + \frac{x^2}{\sqrt{25 - x^2}}$

THINGS TO DO AND THINK ABOUT

1   By expressing $y = \tan^{-1} x$ as $\tan y = x \left(0 \leqslant y < \frac{\pi}{2}\right)$, obtain $\frac{dy}{dx}$ in terms of x only.   3

2   The graph of a function given by $y = \cos^{-1}(e^x - 2x)$ is defined on a suitable domain. Find the x-coordinate of any stationary points in the interval $0 \leqslant x < e$.   4

3   Differentiate $f(x) = \sin^{-1}(4x)$, where $-\frac{1}{4} < x < \frac{1}{4}$.   2

FURTHER DIFFERENTIATION 2 MAC

IMPLICIT DIFFERENTIATION

An equation of the form $f(x, y) = 0$ defines a curve in the xy-plane. This may define y as a function of x, or more than one function of x.

For example, the equation $x^2 + y^2 - 4 = 0$ defines a circle with centre the origin and radius 2.

It defines two functions, $y = +\sqrt{4 - x^2}$ and $y = -\sqrt{4 - x^2}$ for $-2 \leqslant x \leqslant 2$.

Given $f(x, y) = 0$, to obtain $\frac{dy}{dx}$ we use implicit differentiation.

DON'T FORGET

If y is a function of x, then $\frac{d}{dx}[g(y)] = g'(y)\frac{dy}{dx}$, i.e. the chain rule.

Example: 2.13

The equation $x^4 + x^2y^2 + y^4 = 1$ defines y implicitly as a function of x. Obtain $\frac{dy}{dx}$ and $\frac{d^2y}{dx^2}$.

Solution:

Using the product rule and chain rule on the second term, and the chain rule on the third term,

$4x^3 + 2xy^2 + 2x^2y\frac{dy}{dx} + 4y^3\frac{dy}{dx} = 0$.

Gathering terms involving $\frac{dy}{dx}$ gives:

$\frac{dy}{dx}(2x^2y + 4y^3) = -4x^3 - 2xy^2$.

Cancelling a factor of 2 and dividing gives:

$\frac{dy}{dx} = -\frac{2x^3 + xy^2}{2y^3 + x^2y}$ ———— (Eqn 1)

Finding the second derivative gives:

$\frac{d^2y}{dx^2} = -\frac{(2y^3 + x^2y)\left(6x^2 + y^2 + 2xy\frac{dy}{dx}\right) - (2x^3 + xy^2)\left(6y^2\frac{dy}{dx} + 2xy + x^2\frac{dy}{dx}\right)}{(2y^3 + x^2y)^2}$

DON'T FORGET

To differentiate terms such as xy, use the product rule.

Unless you are specifically asked for an expression for $\frac{d^2y}{dx^2}$ in terms of x and y, do not replace $\frac{dy}{dx}$ in this expression by its result from Eqn 1 above. The most likely information you will need from the second derivative is its numerical value for a given (x, y), so simply substitute for x and y in each of the separate derivative expressions.

EQUATIONS OF TANGENTS TO CURVES DEFINED IMPLICITLY

Example: 2.14

Obtain an equation for the tangent at the point $(1, 2)$ on the curve defined by:

$xy^3 - 2x^2y^2 + x^4 - 1 = 0$.

Solution:

$\frac{d}{dx}(xy^3) - \frac{d}{dx}(2x^2y^2) + 4x^3 = 0$, so:

$y^3 + 3xy^2\frac{dy}{dx} - 4xy^2 - 4x^2y\frac{dy}{dx} + 4x^3 = 0$. ——— (Eqn 1)

Let the gradient of the tangent at the point $(1, 2)$ be m.

Then $8 + 12m - 16 - 8m + 4 = 0$, giving $m = 1$, and an equation for the tangent at $(1, 2)$ is $y = x + 1$.

Alternatively to find m, the gradient, rearrange Eqn 1 to get:

$\frac{dy}{dx} = \frac{4xy^2 - 4x^3 - y^3}{3xy^2 - 4x^2y}$ and substitute $x = 1$, $y = 2$

LOGARITHMIC DIFFERENTIATION

Logarithmic differentiation refers to the process of first taking logs (to base e), then differentiating the result of this.

Example: 2.15

Obtain $\frac{d}{dx}(x^{\cos x})$, where $x > 0$.

Solution:

Let $y = x^{\cos x} \Rightarrow \ln y = \cos x \ln x$, so:

$\frac{1}{y}\frac{dy}{dx} = \frac{\cos x}{x} - \sin x \ln x$, which gives:

$\frac{dy}{dx} = x^{(\cos x)-1} \cos x - x^{\cos x} \sin x \ln x$.

Example: 2.16

Obtain $\frac{dy}{dx}$ when $y = \sqrt{\frac{1 + x^2}{1 - x^2}}$, where $-1 < x < 1$.

Solution:

$\ln y = \frac{1}{2}\ln(1 + x^2) - \frac{1}{2}\ln(1 - x^2)$, so:

$\frac{1}{y}\frac{dy}{dx} = \frac{x}{1 + x^2} + \frac{x}{1 - x^2} = \frac{2x}{(1 + x^2)(1 - x^2)}$

$\frac{dy}{dx} = \frac{2xy}{(1 + x^2)(1 - x^2)} = \frac{2x\sqrt{1 + x^2}}{(1 + x^2)(1 - x^2)\sqrt{1 - x^2}}$

$= \frac{2x}{(1 + x^2)^{1/2}(1 - x^2)^{3/2}}.$

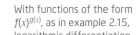

DON'T FORGET

With functions of the form $f(x)^{g(x)}$, as in example 2.15, logarithmic differentiation **must** be used.

DON'T FORGET

If the derivative of a function appears to involve several combinations of the product, quotient or chain rules, then logarithmic differentiation may be best.

THINGS TO DO AND THINK ABOUT

1   Given $xy + y^2 + 4 = 0$, find

   (a) $\frac{dy}{dx}$ 3

   (b) $\frac{d^2y}{dx^2}$ 4

2   The equation $y^2 + 3xy = x^2 - 3$ defines a curve through the point $P(-1, 1)$.
    Obtain the gradient of the tangent to the curve at P.   4

3   Find $\frac{dy}{dx}$ for the function:
    $e^{xy} = y \sin x$ 4

4   Use logarithmic differentiation to obtain the derivative of:
    $y = \frac{(2x - 1)^3}{(x - 2)^2}$ in terms of x, $x > 2$, expressing your answer as a single algebraic fraction.   4

5   Obtain the derivatives of the following functions:

   (a) $f(x) = \exp(-\cos 3x)$ 3

   (b) $y = 2^{x^3 + x}$.   3

ONLINE

Learn more about this topic at www.brightredbooks.net

ONLINE TEST

Test yourself on differentiation at www.brightredbooks.net

FURTHER DIFFERENTIATION 3 MAC

PARAMETRIC DIFFERENTIATION

Equations of the form $x = f(t)$, $y = g(t)$, where t takes values in some interval, describe a curve in the xy-plane. This curve may define a single function y of x, or more than one function. These equations are called **parametric equations**.

For example, $x = t + 1$, $y = t^2$, where t is any real number. Here it is easy to eliminate t to give $y = (x - 1)^2$. However, in the examination, you shouldn't eliminate unless asked to.

We often need to obtain $\frac{dy}{dx}$ (and sometimes $\frac{d^2y}{dx^2}$) when x and y are given parametrically.

Example: 2.17

Obtain $\frac{dy}{dx}$ and $\frac{d^2y}{dx^2}$ given that $x = t^3 + t$, $y = t^2 + 1$.

Solution:

$\frac{dx}{dt} = 3t^2 + 1$, $\frac{dy}{dt} = 2t$

So, $\frac{dy}{dx} = \frac{2t}{3t^2 + 1}$

$\frac{d^2y}{dx^2} = \frac{dy}{dx}\left(\frac{dy}{dx}\right) = \frac{d}{dt}\left(\frac{2t}{3t^2 + 1}\right)\frac{dt}{dx}$

$= \frac{2 - 6t^2}{(3t^2 + 1)^2} \times \frac{1}{3t^2 + 1} = \frac{2 - 6t^2}{(3t^2 + 1)^3}$

RELATED RATES OF CHANGE

Example: 2.18

A particle has its position defined by $x = \frac{1}{2}t^2$; $y = \frac{1}{3}\sqrt{(t + 1)^3}$.
Find expressions, in terms of t, for
(a) the velocity of the particle in the y-direction
(b) the acceleration of the particle in the y-direction
(c) the speed of the particle.

Solution:

(a) $v_y = \frac{dy}{dt} = \frac{d}{dt}\left(\frac{1}{3}(2t + 1)^{\frac{3}{2}}\right) = \frac{3}{2} \times \frac{1}{3}(2t + 1)^{\frac{1}{2}} \times 2 = \sqrt{2t + 1}$

(b) $a_y = \frac{d}{dt}(v_y) = \frac{d}{dt}(2t + 1)^{\frac{1}{2}} = \frac{1}{2}(2t + 1)^{-\frac{1}{2}} \times 2 = \frac{1}{\sqrt{2t + 1}}$

(c) Speed $= \sqrt{\left(\frac{dx}{dt}\right)^2 + \left(\frac{dy}{dt}\right)^2}$

$= \sqrt{(t)^2 + \left(\sqrt{2t + 1}\right)^2}$

$= \sqrt{t^2 + 2t + 1}$

$= \sqrt{(t + 1)^2}$

$= t + 1$

You might be given a problem involving several related variables. In problems relating to rates of change, one of these variables will be time (t). Using differentiation, you can find related rates of change.

For example, given $\frac{dV}{dt}$ and $\frac{dr}{dt}$, look for a relationship between V and r to get:

$\frac{dV}{dt} = \frac{dV}{dr} \cdot \frac{dr}{dt}$

contd

Example: 2.19

A metal cylinder is rolled in a steel press in such a way that it keeps a cylindrical shape, getting longer and thinner, and its volume, V, remains constant. Before rolling, it has radius 2 cm and length 10 cm. The radius is decreasing at a constant rate of 0·2 cm s^{-1}. At what rate is the length increasing when the radius is 1 cm?

Solution:

If the radius and length at any time are r and l,

$\frac{dr}{dt} = -0 \cdot 2$ and $V = \pi r^2 l = \pi 2^2 10 = 40\pi$, so:

$\pi r^2 l = 40\pi$, hence $l = \frac{40}{r^2}$ and:

$\frac{dl}{dt} = \frac{d}{dt}\left(\frac{40}{r^2}\right) = \frac{80}{r^3} \frac{dr}{dt} = \frac{80}{r^3} \times (-0 \cdot 2) = \frac{16}{r^3}$.

So, when the radius is 1 cm, the length is increasing at 16 cm s^{-1}.

DON'T FORGET

Take care with units.

Example: 2.20

Water is being pumped into a conical tank radius 3 m, height 5 m.
Initially, the tank is empty. The water is pumped at a rate of 3 m^3 per minute.
Find the rate at which the water level is rising after 5 minutes.

Solution:

We need to find $\frac{dh}{dt}$ given that $\frac{dV}{dt} = 3$.

$V_t = \frac{1}{3}\pi r^2 h$ $r_t = \frac{3}{5}h$.

$\therefore V_t = \frac{1}{3}\pi \left(\frac{3}{5}h\right)^2 h = \frac{1}{3} \times \frac{9}{25}\pi h^3 = \frac{3\pi h^3}{25}$

$\therefore \frac{dV_t}{dh} = \frac{9\pi h^2}{25}$ $\therefore \frac{dh}{dV_t} = \frac{25}{9\pi h^2}$

$\therefore \frac{dh}{dt} = \frac{dV}{dt} \times \frac{dh}{dV_t} = \frac{3 \times 25}{9\pi h^2} = \frac{25}{3\pi h^2}$

$\therefore$ when $t = 5$, $V = 15 \text{m}^3$ $\therefore 15 = \frac{3\pi h^3}{25}$

$\therefore h = \sqrt[3]{\frac{125}{\pi}} = 3 \cdot 4139... \text{m}$

$\therefore \frac{dh}{dt} = \frac{25}{3\pi(3 \cdot 4139...)^2} = 0 \cdot 228 \text{ m/min}$

THINGS TO DO AND THINK ABOUT

1 A hollow circular cone (see diagram) is fixed in a vertical position and filled at a constant rate of 2 cm^3 per second. Calculate the rate at which the depth, x, is increasing when $x = 3$.

5

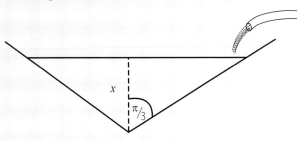

2 Given $x = \tan \theta$, $y = \sin \theta$, show that $\frac{dy}{dx} = \cos^3 \theta$.

3

ONLINE

Find out more about derivatives at www.brightredbooks.net

ONLINE TEST

Test yourself on differentiation at www.brightredbooks.net

FUNCTIONS

PROPERTIES OF FUNCTIONS AAC

A function is a mathematical relationship such that each value of x in the **domain** is associated with a value of y in the **range**.

We write $y = f(x)$, and say that y is the value of f at x. The set of all values y obtained this way is called the **range** of f.

Example: 3.1

$f(x) = -\sqrt{x}$ for all $x > 0$. Describe the domain and range.

Solution:

Here the domain is the set of all positive real numbers, while the range is the set of all negative real numbers.

In Advanced Higher, the function rule is usually given by a formula, or combination of formulae. If the domain is not specified, it can be assumed to be $\mathbb{R}$ (the set of all real numbers), or the largest subset of $\mathbb{R}$ for which the rule makes sense. The range is not usually specified, but you could be asked to obtain it.

EVEN AND ODD FUNCTIONS

If $f(x) = f(-x)$ for all x in the domain of f, then f is said to be **even**. The graph of f is symmetrical, with $x = 0$ the axis of symmetry.

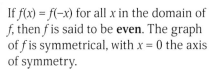

If $f(x) = -f(-x)$ for all x in the domain of f, then f is said to be **odd**. The graph of f has **half-turn** symmetry about the origin.

For polynomial functions:

All ODD powers → ODD function: e.g. $x^3 + 2x$.

All EVEN powers → EVEN function: e.g. $x^6 - 2x^2 + 3$.

Example: 3.2

$f(x) = \cos x + x^2$: f is even.

$g(x) = x + \sin x$: g is odd.

$h(x) = x^3 + x^2 + x$: h is neither even nor odd, but it is the sum of an even function (the function x^2) and an odd function (the function $x^3 + x$).

The product or quotient of two ODD functions produces an EVEN function.

Similarly, the product or quotient of two EVEN functions is an EVEN function.

Also, the product or quotient of an ODD and an EVEN function is ODD.

INVERSE FUNCTIONS

If $y = f(x)$, and we plot the points (x, y), we get the graph of f.

If the function f has an inverse, every horizontal line meets the graph $y = f(x)$ once at most. The graph of f^{-1} is obtained by reflecting the graph of f in the line $y = x$.

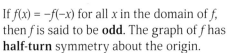

ASYMPTOTES

Consider the curve $y = x - 2 + \frac{1}{x^2 + 1}$ where x is real.

As x gets large, $x^2 + 1$ gets very large and $\frac{1}{x^2 + 1}$ gets very small.

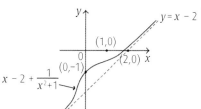

So, the value of y gets closer and closer to the value $x - 2$, which means that, for large x, points on the curve are close to the line $y = x - 2$.

We say that $y = x - 2$ is an **asymptote** of the curve.

Some graphs have **vertical asymptotes**. If $f(a)$ is not defined, and $f(x)$ becomes larger and larger as $x \to a$, then we say that $x = a$ is a vertical asymptote of the curve $y = f(x)$.

Example: 3.3

Given $y = \frac{x + 3}{x + 2}$, we can divide to get $y = 1 + \frac{1}{x + 2}$. The curve defined by this equation has two asymptotes: the **horizontal asymptote** $y = 1$ and the **vertical asymptote** $x = -2$.

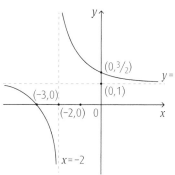

Most functions met at Advanced Higher are continuous, except at isolated points such as vertical asymptotes. An intuitive idea of continuity is that the graph of a continuous function has no breaks in it.

💭 THINGS TO DO AND THINK ABOUT

1 Determine whether the function $f(x) = x^3 \tan x$ is odd, even or neither. Justify your answer. **3**

2 Part of the graph of a function, f, is shown:
   - f is an odd function
   - $x = 2$ is an asymptote of the graph
   - $y = x$ is also an asymptote.

   (a) Copy the diagram and complete the graph. **3**

   (b) State the equation of the new asymptote. **1**

   (c) Sketch $y = |f(x)|$ showing all relevant information. **3**

3 Part of the graph $y = f(x)$ is shown, where the dotted lines indicate asymptotes. Sketch the graph of $y = -f(x + 1)$, showing its asymptotes. Write down the equations of the asymptotes. **4**

   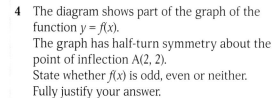

4 The diagram shows part of the graph of the function $y = f(x)$.
   The graph has half-turn symmetry about the point of inflection A(2, 2).
   State whether $f(x)$ is odd, even or neither.
   Fully justify your answer. **3**

   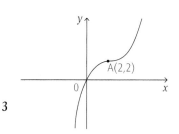

CRITICAL POINTS 1

DON'T FORGET

Critical points are:
- end points
- "kinks"
- stationary points.

DON'T FORGET

Stationary points may be local maxima/minima, or points of inflexion.

OVERVIEW

Any point $(a, f(a))$ on the graph of f for which either $f'(a) = 0$ or $f'(a)$ does not exist is called a **critical point** of f.

If $f'(a) = 0$, the point $(a, f(a))$ is called a **stationary** point (S.P.). S.P.s are also critical points.

Example: 3.4

Consider the function defined for $-1 \le x \le 1$ by $f(x) = |x|$. The graph of f shows that f is continuous; $f'(0)$ does not exist, since it changes **instantaneously** from a value of -1 when $x < 0$ to a value of $+1$ when $x > 0$. (In other words, the change in $f'(x)$ value is not 'smooth' – visually, there is a 'kink' at the origin.) So, $(0, 0)$ is a critical point. The points $(-1, 1)$ and $(1, 1)$ are also critical points, because $x = -1$ and $x = 1$ are the end points of this function.

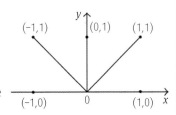

A point (x_0, y_0) on the graph $y = f(x)$ where the curve changes from concave up, when $f''(x) > 0$, to concave down, when $f''(x) < 0$, or vice versa, is called a **point of inflexion**. At such a point, $f''(x_0) = 0$.

Testing for a point of inflexion requires that you find the x-value for which $f''(x) = 0$ and check that the value of $f''(x)$ changes sign to the left and right of this x-value.

1. Rising stationary point of inflexion

   $f'(a) = 0$ AND $f''(a) = 0$

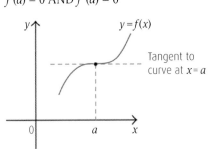

2. Falling stationary point of inflexion

   $f'(a) = 0$ AND $f''(a) = 0$

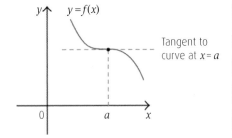

3. Sloping increasing point of inflexion

   $f'(a) > 0$ AND $f''(a) = 0$

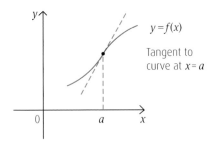

4. Sloping decreasing point of inflexion

   $f'(a) < 0$ AND $f''(a) = 0$

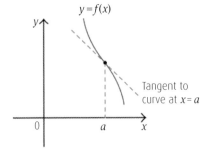

Thus, the gradient need **not** be 0 at a point of inflexion.

contd

Example: 3.5

Consider $f(x) = x^3$ for real x.

$f''(x) = 6x$, so $(0, 0)$ is a point of inflexion.

The curve changes from concave down for $x < 0$ to concave up for $x > 0$.

Critical points given by solving $f'(x) = 0$ fall into one of three groups:

- local or global **maxima**
- local or global **minima**
- **points of inflexion**.

Maxima and minima can be determined by looking at the sign of f' near the critical point (nature table). Alternatively, if f'' is easy to obtain, look at the sign of f'' at the critical point.

- $f''(a) > 0 \Rightarrow$ minimum at $x = a$
- $f''(a) < 0 \Rightarrow$ maximum at $x = a$
- $f''(a) = 0 \Rightarrow$ stationary point or a point of inflexion.

$f''(x) > 0$ means curve is concave up, so if $f'(a) = 0$ and $f''(a) > 0$ we have

$f''(x) < 0$ means curve is concave down, so if $f'(a) = 0$ and $f''(a) < 0$ we have

$f'(a) = 0 = f''(a)$ can be a point of inflexion or a stationary point.

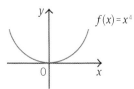

For $f(x) = x^4$, $f'(0) = f''(0) = 0$ with minimum at $x = 0$.

THINGS TO DO AND THINK ABOUT

1 Part of the graph of the function $y = f(x)$ is shown below. Sketch $y = |f(x)|$, stating the coordinates of all critical points. **4**

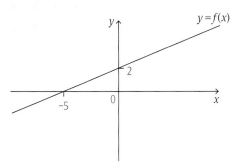

2 Part of the graph of the linear function $y = f(x)$ is shown below. Sketch $y = |2 - f(x)|$, stating the coordinates of any critical points. **4**

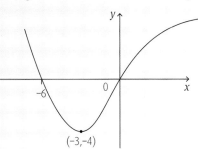

CRITICAL POINTS 2 AAC

Example: 3.6

$f(x) = \frac{x^2 - 1}{x^2 + 1}$. Find the asymptotes and points of inflexion, then sketch the graph of f.

Solution:

Dividing gives $f(x) = 1 - \frac{2}{x^2 + 1}$, which makes it easier to obtain f'.

$f'(x) = \frac{4x}{(x^2 + 1)^2}$ and $f''(x) = \frac{4 - 12x^2}{(x^2 + 1)^3}$

Note that the graph $y = f(x)$ is symmetrical about the y-axis because f is even: $(f(x) = f(-x))$.

There are no vertical asymptotes since $x^2 + 1 \neq 0$, but because $f(x) \to 1$ as $x \to \pm \infty$, $y = 1$ is a horizontal asymptote.

Stationary points are given by $f'(x) = 0 \Rightarrow x = 0$, so $(0, -1)$ is a stationary point.

As $f''(0) > 0$, this gives a global minimum. Alternatively, you can easily check that the gradient as x passes through $x = 0$ changes: ↘_↗.

To look for the existence of points of inflexion, we solve $f''(x) = 0$ to find that $x = \pm \frac{1}{\sqrt{3}}$.

In this case, we only need to investigate what happens for the positive value of x, because the graph is symmetrical about the y-axis.

$f''\left(\frac{1}{\sqrt{3}}\right) = 0$ but $f'\left(\frac{1}{\sqrt{3}}\right) \neq 0$, so this gives a sloping point of inflexion.

To determine whether this is a rising/increasing point of inflexion or falling/decreasing point of inflexion, use either nature table for $f(x)$ **or** nature table for $f''(x)$ (i.e. concavity of curve).

- $f'\left(-\frac{1}{\sqrt{3}}\right) < 0 \Rightarrow$ falling point of inflexion.
- $f'\left(-\frac{1}{\sqrt{3}}\right) > 0 \Rightarrow$ rising point of inflexion

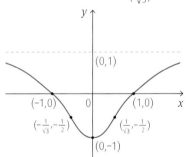

Example: 3.7

$f(x) = x^4 - 4x^3$. Find the points of inflexion and sketch the graph.

Solution:

$f'(x) = 4x^3 - 12x^2$ and $f''(x) = 12x^2 - 24x$

$f'(x) = 0 \Rightarrow x = 0$ and $x = 3$.

Using a nature table for f', we get:

so there is a horizontal point of inflexion at $x = 0$, and a (global) minimum at $x = 3$.

→	0	→	3	→
−	0	−	0	+
\	−	\	−	/

The positions of points of inflexion are given by solving $f''(x) = 12x^2 - 24x = 0$, giving $x = 0$ (already obtained) and $x = 2$. As $f''(x) < 0$ for $x < 2$ and $f''(x) > 0$ for $x > 2$, the curve changes from concave down to concave up at $x = 2$.

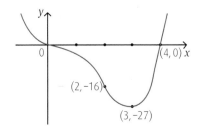

Note that in this example the second derivative test for the nature of any turning points is not completely sufficient.

Because $f'(3) = 0$ and $f''(3) > 0$, we can conclude that $x = 3$ gives a minimum. However, because $f'(0) = f''(0) = 0$, we have to resort to a nature table for $x = 0$.

contd

Steps for sketching functions

- Is there any obvious symmetry (odd, even, neither)?
- Find, if possible, where the curve cuts the x- and y-axes.
- Examine the behaviour of the function as $x \to \pm \infty$, including horizontal and oblique asymptotes.
- Investigate any values of x for which the function is undefined (vertical asymptotes).
- Investigate critical points.

THINGS TO DO AND THINK ABOUT

1 The function f is defined by $f(x) = \dfrac{x^2 + 3x}{x + 1}$ $(x \neq -1)$.

    (a) Obtain equations of the asymptotes of the graph of f.   **3**

    (b) Show that the graph is always increasing.   **3**

    (c) Sketch the graph of f, showing all important features.   **2**

2 Part of the graph of the function $f(x) = \dfrac{kx + 5}{kx - 2}$; $k > 1$, $x \neq \dfrac{2}{k}$ is shown below.

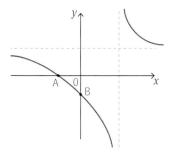

 ONLINE

Find out more at
www.brightredbooks.net

    (a) Obtain the coordinates of points A and B.   **3**

    (b) State the equations of the two asymptotes.   **2**

    (c) Show that the graph of $f(x)$ has no stationary values.   **3**

 VIDEO LINK

Check out the clip at
www.brightredbooks.net
for more on this topic.

3 The function f is defined by $f(x) = \dfrac{x - 3}{x + 2}$, $x \neq -2$, and the diagram shows part of its graph.

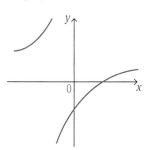

 ONLINE TEST

Test yourself on
critical points at
www.brightredbooks.net

    (a) Obtain algebraically the asymptotes of the graph of f.   **3**

    (b) Prove that f has no stationary values.   **2**

    (c) Does the graph of f have any points of inflexion? Justify your answer.   **2**

    (d) Sketch the graph of the inverse function, f^{-1}. State the asymptotes and the domain of f^{-1}.   **3**

RELATED GRAPHS

OVERVIEW

If you are given a graph $y = f(x)$ (sometimes only as a diagram without a formula for f), there are several related graphs which you could be asked to sketch. In what follows, k is a constant.

(i) $y = kf(x)$

(ii) $y = f(kx)$

(iii) $y = f(x) + k$

(iv) $y = f(x + k)$

(v) $y = |f(x)|$

(vi) $y = f'(x)$

(vii) $f^{-1}(x)$

(viii) a combination of (i) to (viii)

Example: 3.8

The diagram shows the graph of a function f.

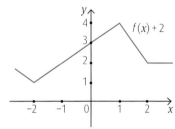

Sketch the graphs of $2f(x)$, $f(2x)$, $f(x) + 2$, $f(x + 2)$, $|f(x)|$.

Solution:

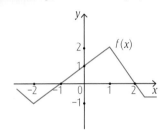

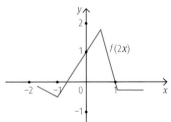

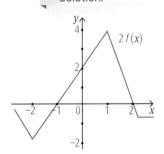

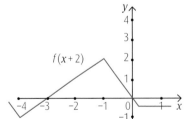

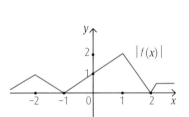

Example: 3.9

Part of the graph of $y = f(x)$ is shown.

Sketch: (a) $|f^{-1}(x)|$

       (b) $|f(2x) - 6|$

       (c) $f^{-1}(x + 2)$

       (d) $f'(3x)$

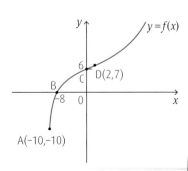

contd

Solution:

(a)

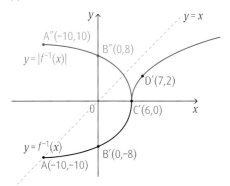

(b)

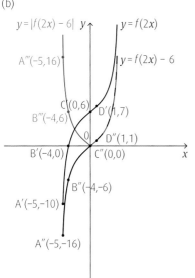

(c)

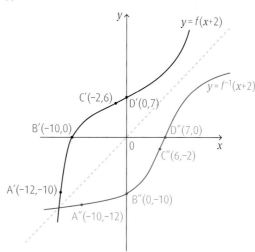

(d)

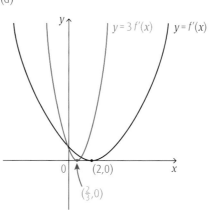

🗨️ THINGS TO DO AND THINK ABOUT

Part of the graph of the function $y = f(x)$ is shown below.

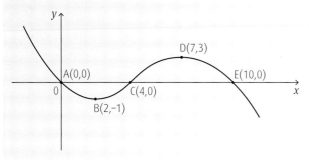

Sketch: (a) $y = -f'(x)$

       (b) $y = f'(-x)$

3

3

GRAPHS OF TRIGONOMETRIC FUNCTIONS

OVERVIEW

Example: 3.10

Part of the graph of the straight-line function $f(x)$ is shown.

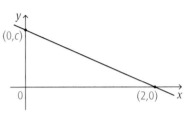

(a) Sketch the graph of $f^{-1}(x)$, showing points of intersection with the axes.

(b) State the value of k for which $f(x) + k$ is an odd function.

(c) Find the value of h for which $|f(x + h)|$ is an even function.

Solution:

(a)

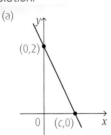

(b) $y = f(x) - c$ is odd. $\therefore k = -c$

(c) $y = |f(x + 2)|$ is even. $\therefore h = 2$

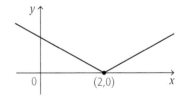

Example: 3.11

The function $f(x)$ is defined for all $x \geq 0$. The graph of $y = f(x)$ intersects the y-axis at $(0, c)$, where $0 < c < 5$. The graph of the function and its asymptote, $y = x - 5$, is shown below.

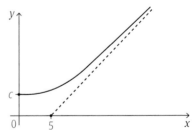

(a) Copy the above diagram.
On the same diagram, sketch the graph of $y = f^{-1}(x)$.
Clearly show any points of intersection and any asymptotes.

(b) What is the equation of the asymptote of the graph of $y = f(x + 2)$?

(c) Why does your diagram show that the equation $x = f(f(x))$ has at least one solution?

Solution:

(a)

(b) Assymptote passes through $(3, 0)$ so equation is $y = x - 3$

(c) $f^{-1}(x) = f(x)$ at intersection of curves on line $y = x$.
So $x = f(f(x))$

contd

You should be able to sketch the graphs of the three basic functions sin x, cos x and tan x, their inverse functions and simple related functions, e.g. k cos x and sin kx.

sin x
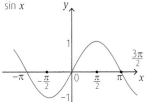

cos x
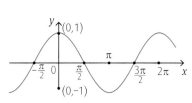

tan x
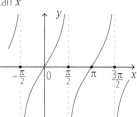

sin^{-1}x
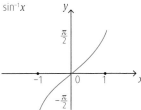

cos^{-1}x
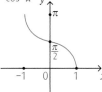

tan^{-1}x
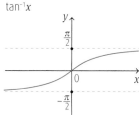

Example: 3.12

Sketch the graph of sec x over the interval $-\pi \leq x \leq \pi$, and the graph of sec^{-1} x.

Solution:

There are three things to note about the graph of sec x before you start:
1. The range is in two parts, $y \geq 1$ and $y \leq -1$.
2. There are vertical asymptotes at $x = \pm \frac{\pi}{2}$.
3. The graph is symmetrical about $x = 0$.

sec x
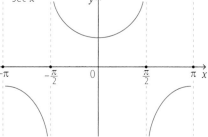

ONLINE

Learn more about this topic by following the link at www.brightredbooks.net

To define the inverse function, we must restrict the domain to $0 \leq x \leq \pi$, which gives the graph of sec^{-1} x as shown in the graph here:

sec^{-1}x
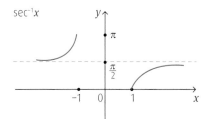

VIDEO LINK

Check out the clip about graphs of inverse trig functions at www.brightredbooks.net

THINGS TO DO AND THINK ABOUT

Now practice the skills shown in example 3·12 by sketching the graph of cosec x over the interval $-\pi \leq x \leq \pi$.

ONLINE TEST

Test yourself on related graphs at www.brightredbooks.net

INTEGRATION

BASIC INTEGRATION AND INTEGRATION BY SUBSTITUTION MAC

OVERVIEW

The indefinite integral of a given function $f(x)$, denoted by $\int f(x)dx$, is a function $F(x)$ which satisfies $\frac{dF}{dx} = f$. This notation was introduced by Leibniz in 1675. The $\int$ symbol is an elongated S, meaning 'sum', and arose in the calculation of areas.

The definite integral: $\int_a^b f(x)dx = F(b) - F(a) = [F(x)]_a^b$

An indefinite integral is a function of x, whereas a definite integral is a **number**.

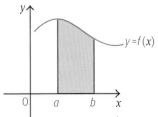

Area of shaded region is $\int_0^b f(x)dx$

Example: 4.1 (reminders from Higher Maths)

$\int \frac{1}{x^3}dx = \frac{-1}{2x^2} + c$, where c is an arbitrary constant. $\int_1^2 \frac{1}{x^3}dx = \left[\frac{-1}{2x^2}\right]_1^2 = \frac{-1}{8} + \frac{1}{2} = \frac{3}{8}$

DON'T FORGET

$\int x^p dx = \frac{x^{p+1}}{p+1} + c$ for any value of p, including negative and fractional values, **except** the value $p = -1$.

INTEGRATION BY SUBSTITUTION

Many integrals can be simplified by replacing x by another carefully chosen variable. A simple case is $u = ax + b$, where a and b are constants, which gives:

$\int f(ax + b)dx = \frac{1}{a}\int f(u)du.$

Combining this with the table of standard derivatives (see the table in Chapter 2, page 16) gives the following useful standard integrals:

DON'T FORGET

$\int \frac{1}{x}dx = \ln|x| + c$

Standard integrals

As with differentiation, many of the following integrals now appear in the formulae sheet, but you should try to learn them so that you recognise ones that are similar but not identical.

DON'T FORGET

Make sure you distinguish between:
$\int \frac{1}{a^2 + x^2}dx$ and
$\int \frac{x}{a^2 + x^2}dx$
$= \frac{1}{2}\ln(a^2 + x^2) + c$

DON'T FORGET

Make sure you distinguish between:
$\int \frac{1}{\sqrt{a^2 - x^2}}dx$ and
$\int \frac{x}{\sqrt{a^2 - x^2}}dx$
$= -\sqrt{a^2 - x^2} + c$

$f(x)$	$\int f(x)dx$
e^x	$e^x + c$
$\frac{1}{x}$	$\ln|x| + c$ (or $\log_e|x| + c$)
$\tan x$	$-\ln|\cos x| + c$ (or $\ln|\sec x| + c$)
$\cot x$	$\ln|\sin x| + c$
$\sin^2 x$	$\frac{1}{2}x - \frac{1}{4}\sin 2x + c$
$\cos^2 x$	$\frac{1}{2}x + \frac{1}{4}\sin 2x + c$
$\sec^2 x$	$\tan x + c$
$\sec x \tan x$	$\sec x + c$
$\csc^2 x$	$-\cot x + c$
$\frac{1}{\sqrt{1 - x^2}}$	$\sin^{-1} x + c$
$\frac{1}{1 + x^2}$	$\tan^{-1} x + c$
$\frac{1}{\sqrt{a^2 - x^2}}$	$\sin^{-1}\left(\frac{x}{a}\right) + c$
$\frac{1}{a^2 + x^2}$	$\frac{1}{a}\tan^{-1}\left(\frac{x}{a}\right) + c$

contd

Example: 4.2

$$\int_0^2 \frac{1}{x^2 + 4}\, dx = \frac{1}{2}\left[\tan^{-1}\left(\frac{x}{2}\right)\right]_0^2 = \frac{1}{2}\tan^{-1}(1) = \frac{\pi}{8}$$

When the substitution is more complicated, you will usually be given the substitution to be used.

$\int \frac{f'(x)}{f(x)}\, dx = \ln|f(x)| + c$ is an integral you should remember but it may also be obtained by the substitution $u = f(x)$.

Using the differential notation, $du = f'(x)dx$, giving:

$$\int \frac{f'(x)}{f(x)}\, dx = \int \frac{1}{u}\, du = \ln|u| + c = \ln|f(x)| + c.$$

Example: 4.3

Use the substitution $x = 2\tan u$ to evaluate $\int_0^2 (4 + x^2)^{-\frac{3}{2}}\, dx$.

Solution:

Differentiating the substitution $x = 2\tan u$ gives:

$dx = 2\sec^2 u\, du$

when $x = 0$, $u = 0$, and when $x = 2$, $u = \frac{\pi}{4}$.

Now we need to simplify the function to be integrated, so we do this separately before continuing with the integration:

$$(4 + x^2)^{-\frac{3}{2}} = (4\sec^2 u)^{-\frac{3}{2}} = \frac{1}{8\sec^3 u} = \frac{\cos^3 u}{8}$$

So, we now have:

$$\int_0^2 (4 + x^2)^{-\frac{3}{2}}\, dx = \frac{1}{8}\int_0^{\frac{\pi}{4}} (\cos^3 u)\, 2\sec^2 u\, du = \frac{1}{4}\int_0^{\frac{\pi}{4}} \cos u\, du = \frac{1}{4}\left[\sin u\right]_0^{\frac{\pi}{4}}$$

$$= \frac{1}{4}\left[\sin \frac{\pi}{4} - \sin 0\right] = \frac{1}{4\sqrt{2}} = \frac{\sqrt{2}}{8}.$$

Follow these steps when using integration by substitution:

- Obtain dx as (something)du; the 'something' can be left in terms of x here if necessary.

- Change the limits (for a definite integral).

- Substitute for **ALL** occurrences of x and dx.

- Simplify, and make sure the integrand is now completely in terms of u.

- Integrate.

- For an indefinite integral, remember to change back to the variable x.

THINGS TO DO AND THINK ABOUT

1. Use the substitution $x = (u - 1)^2$ to obtain $\int \frac{1}{(1 + \sqrt{x})^3}\, dx$.   5

2. (a) Use the substitution $u = 1 + x^2$ to obtain   5
   $$\int_0^1 \frac{x^3}{(1 + x^2)^4}\, dx$$
   (b) A solid is formed by rotating the curve $y = \frac{x^{\frac{3}{2}}}{(1 + x^2)^2}$ between $x = 0$ and $x = 1$ through 360° about the x-axis. Write down the volume of this solid.   1

3. Use the substitution $x = 2\sin\theta$ to obtain the exact value of $\int_0^{\sqrt{2}} \frac{x^2}{\sqrt{4 - x^2}}\, dx$.   6
   (Note that $\cos 2A = 1 - 2\sin^2 A$.)

DON'T FORGET

$\int \frac{f'(x)}{f(x)}\, dx = \ln|f(x)| + c.$

DON'T FORGET

For indefinite integrals, remember to change back to the variable x. For definite integrals, change the limits to correspond to the new variable. Alternatively, change the variable back before evaluating.

DON'T FORGET

Where possible and sensible, leave your answer as an exact value.

VIDEO LINK

Learn more about this topic by watching the clip at www.brightredbooks.net

ONLINE

Investigate integration by substitution further at www.brightredbooks.net

ONLINE TEST

Test yourself on integration at www.brightredbooks.net

FURTHER INTEGRATION MAC

PARTIAL FRACTIONS

Integrals of the form $\int \frac{p(x)}{q(x)} dx$, where p and q are polynomials, can be obtained by expressing $\frac{p(x)}{q(x)}$ in partial fractions. See pp 10–15 for details about partial fractions.

DON'T FORGET

Partial fractions can be useful for differentiation as well as integration.

Example: 4.4

Obtain $\int \frac{x^2}{x^2 - x - 2} dx$.

Solution:

Using long division gives: $\frac{x^2}{x^2 - x - 2} = 1 + \frac{x + 2}{x^2 - x - 2}$.

We now set $\frac{x + 2}{x^2 - x - 2} = \frac{A}{x + 1} + \frac{B}{x - 2}$ for constants A and B.

This gives $x + 2 = A(x - 2) + B(x + 1)$, from which we get $A = -\frac{1}{3}$, $B = \frac{4}{3}$.

Hence $\int \frac{x^2}{x^2 - x - 2} dx = \int \left(1 + \frac{4/3}{x - 2} - \frac{1/3}{x + 1}\right) dx = x + \frac{4}{3}\ln|x - 2| - \frac{1}{3}\ln|x + 1| + c$.

DON'T FORGET

Remember the modulus sign in log functions.

Example: 4.5

Evaluate $\int_0^2 \frac{x - 1}{(x - 3)^2} dx$.

Solution:

Let $\frac{x - 1}{(x - 3)^2} = \frac{A}{x - 3} + \frac{B}{(x - 3)^2}$.

This gives $x - 1 = A(x - 3) + B$ and hence $A = 1$ and $B = 2$.

Applying this result gives $\int_0^2 \frac{x - 1}{(x - 3)^2} dx = \int_0^2 \left(\frac{1}{x - 3} + \frac{2}{(x - 3)^2}\right) dx$

$= \left[\ln|x - 3| - \frac{2}{x - 3}\right]_0^2 = \frac{4}{3} - \ln 3$.

DON'T FORGET

Note that the modulus sign is absolutely essential here, and its omission would lose marks.

Example: 4.6

Obtain $\int \frac{1}{x^3 + 4x} dx$.

Solution:

We factorise $x^3 + 4x$ to get $x(x^2 + 4)$ and let $\frac{1}{x^3 + 4x} = \frac{A}{x} + \frac{Bx + C}{x^2 + 4}$.

So, $A(x^2 + 4) + x(Bx + C) = 1$.

Setting $x = 0$ gives $A = \frac{1}{4}$.

Equating coefficients of x^2 gives: $A + B = 0 \Rightarrow B = -\frac{1}{4}$.

Equating coefficients of x gives $C = 0$.

Then $\int \frac{1}{x^3 + 4x} dx = \frac{1}{4}\int \left(\frac{1}{x} - \frac{x}{x^2 + 4}\right) dx = \frac{1}{4}\ln|x| - \frac{1}{8}\ln(x^2 + 4) + c$.

DON'T FORGET

The form of the polynomial q in $\frac{p(x)}{q(x)}$ could be any of the following types:
- a quadratic function
- a product of three linear factors
- the product of a linear factor and an irreducible quadratic factor.

AREAS AND VOLUMES AAC

Integration can be used to obtain areas under curves, areas between two curves, and also volumes of revolution.

Example: 4.7

Calculate the area in the first quadrant bounded by the x-axis, the y-axis and the curve with equation $y = 6 + x - x^2$.

Solution:

$x^2 - x - 6 = (x + 2)(x - 3)$, so the curve crosses the x-axis at $x = 3$, since in the first quadrant $x > 0$.

Required area $= \int_0^3 (6 + x - x^2) dx = \left[6x + \frac{x^2}{2} - \frac{x^3}{3}\right]_0^3 = \frac{27}{2}$ units^3

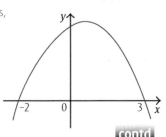

contd

Example: 4.8

Find the volume of revolution formed when the shaded part bounded by the lines $x = 0$, $y = 2$, $y = 2\sqrt{3}$ and $y = \sqrt{\frac{1 - 4x^2}{x^2}}$ is rotated 2π about the y-axis.

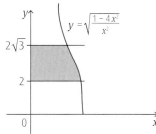

Solution:

$$y = \sqrt{\frac{1 - 4x^2}{x^2}}$$

$$\therefore y^2 = \frac{1 - 4x^2}{x^2}$$

$$\therefore x^2 y^2 + 4x^2 = 1$$

$$\therefore x^2 = \frac{1}{4 + y^2}$$

$$\therefore V = \pi \int_2^{2\sqrt{3}} \frac{1}{4 + y^2} \, dy$$

$$= \pi \left[\frac{1}{2} \tan^{-1}\left(\frac{y}{2}\right)\right]_2^{2\sqrt{3}}$$

$$= \pi \left[\left(\frac{1}{2}\tan^{-1}\sqrt{3}\right) - \left(\frac{1}{2}\tan^{-1}1\right)\right] = \frac{1}{2}\pi \left(\frac{\pi}{3} - \frac{\pi}{4}\right) = \frac{\pi^2}{24} y^3 \text{ units}^3$$

DON'T FORGET

Volume of revolution about x-axis is $\pi \int_a^b y^2 \, dx$ (a and b are values of x).

DON'T FORGET

Volume of revolution about y-axis is $\pi \int_q^p x^2 \, dy$ (p and q are values of y).

Example: 4.9

A solid is formed by rotating the curve $y = e^{-2x}$ between $x = 0$ and $x = 1$ through 360° about the x-axis. Calculate the volume of the solid.

Solution:

The volume is given by $\int_0^1 \pi y^2 \, dx = \pi \int_0^1 e^{-4x} \, dx = \pi \left[\frac{-1}{4}e^{-4x}\right]_0^1 = \frac{\pi}{4}(1 - e^{-4}) \text{ units}^3$

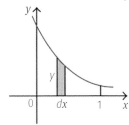

Volume $\int \pi y^2 dx$
Summing gives $\pi \int_0^1 y^2 dx$

ONLINE

Follow the link at www.brightredbooks.net for more on this topic.

VIDEO LINK

Watch the volumes-of-revolution animation at www.brightredbooks.net

THINGS TO DO AND THINK ABOUT

1 **(a)** Express $\dfrac{1}{x^2 + 2x - 8}$ in partial fractions. **2**

(b) Evaluate $\displaystyle\int_0^1 \dfrac{1}{x^2 + 2x - 8} \, dx$. **4**

2 **(a)** Express $\dfrac{1}{x^3 + x}$ in partial fractions. **4**

(b) Obtain a formula for $I(k)$, where $I(k) = \displaystyle\int_1^k \dfrac{1}{x^3 + x} \, dx$, expressing it in the form $\ln\left(\frac{a}{b}\right)$, where a and b depend on k. **4**

3 Find the volume of the solid formed when the area enclosed by the curve $y = x^2 - 1$, the x-axis and the line $x = 2$ is rotated 2π about the x-axis. **5**

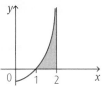

4 Find the volume of the solid formed when the area enclosed by the curve $y = x^4 - 1$, the x-axis, the y-axis and the line $y = 3$ is rotated 2π about the y-axis. **5**

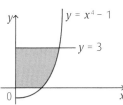

ONLINE TEST

Test yourself on integration at www.brightredbooks.net

INTEGRATION BY PARTS (MAC)

OVERVIEW

Integration by parts is the integration equivalent of the product rule from differentiation. The idea of integration by parts is to produce an integral which is easier to deal with. If you are left with an integral that looks harder than, or as hard as, the integral you started with, then you may not have made the best choice of f and g'. Swapping them over may be all that is required.

The basic result is $\int f(x)g'(x)dx = f(x)g(x) - \int f'(x)g(x)dx$, or the equivalent form:

$\int u\frac{dv}{dx}dx = uv - \int v\frac{du}{dx}dx$ (with a corresponding result for definite integrals).

Its application requires a sensible choice of the functions f and g (or u and v).

The mnemonic LIPET can be useful in making your choice of function for u (the one you are going to differentiate).

L = log
I = inverse trig
P = polynomial
E = exponential
T = trig

Choose u or $f(x)$ in this order.

Example: 4.10

Evaluate $\int_{0}^{2}(x + 1)(x - 2)^3 dx$.

Solution:

Let $f(x) = x + 1$, $g'(x) = (x - 2)^3$ to give:

$f'(x) = 1$; $g(x) = \frac{1}{4}(x - 2)^4$

$\int_{0}^{2}(x + 1)(x - 2)^3 dx = \left[(x + 1)\frac{(x - 2)^4}{4}\right]_0^2 - \frac{1}{4}\int_{0}^{2}(x - 2)^4\,dx$

$\qquad\qquad\qquad = -4 - \frac{1}{20}[(x - 2)^5]_0^2 = -\frac{28}{5}.$

Note that, in the above example, the alternative choice of $g'(x) = x + 1$ and $f(x) = (x - 2)^3$ leads to an integral which requires several integrations by parts.

Example: 4.11 (repeated application)

Obtain $\int x^2 \sin 2x\,dx$.

DON'T FORGET

Sometimes the method may need a second application.

Solution:

$f(x) = x^2 \qquad g(x) = -\frac{1}{2}\cos 2x$
$\quad\downarrow \qquad\qquad\qquad \uparrow$
$f'(x) = 2x \qquad g'(x) = \sin 2x$

A first application gives $\int x^2 \sin 2x\,dx = -\frac{x^2}{2}\cos 2x + \int x \cos 2x\,dx$.

In this second integral, we set $f(x) = x$, $g'(x) = \cos 2x$, which gives:

$\int x \cos 2x\,dx = \frac{x}{2}\sin 2x - \int \frac{\sin 2x}{2}dx = \frac{x}{2}\sin 2x + \frac{1}{4}\cos 2x + c.$

Combining these results, we have:

$\int x^2 \sin 2x\,dx = -\frac{x^2}{2}\cos 2x + \frac{x}{2}\sin 2x + \frac{1}{4}\cos 2x + c.$

contd

For some functions f, $\int f(x)dx$ can be obtained by setting $u = f(x)$, $\frac{dv}{dx} = 1$.

Example: 4.12

Evaluate $\int_0^{1/2} \sin^{-1} 2x \, dx$.

Solution:

Set $u = \sin^{-1} 2x$ and $\frac{dv}{dx} = 1$, so:

$\frac{du}{dx} = \frac{2}{\sqrt{1 - 4x^2}}$ and $v = x$.

Thus $\int_0^{1/2} \sin^{-1} 2x \, dx = [x \sin^{-1} 2x]_0^{1/2} - \int_0^{1/2} \frac{2x}{\sqrt{1 - 4x^2}} \, dx$.

The first term is $\frac{1}{2}\sin^{-1} 1 = \frac{\pi}{4}$.

The second integral can be obtained in various ways.

If you recognise that the numerator is the derivative of $1 - 4x^2$, apart from a constant multiplicative factor, you can guess that the integral must be of the form $k\sqrt{1 - 4x^2}$, and it is then easy to decide what k must be.

Alternatively, make the substitution $u = 4x^2$.

Either method leads to the result:

$\int_0^{1/2} \frac{2x}{\sqrt{1 - 4x^2}} \, dx = \left[-\frac{1}{2}\sqrt{1 - 4x^2}\right]_0^{1/2} = \frac{1}{2}$.

Combining these, we get $\int_0^{1/2} \sin^{-1} 2x \, dx = \frac{\pi}{4} - \frac{1}{2}$.

Example: 4.13 (repeated Integrand)

$\int e^{2x} \cos 3x \, dx$

Solution:

$u = e^{2x}$ $\quad\quad v' = \cos 3x$

$u' = 2e^{2x}$ $\quad\quad v = \frac{1}{3}\sin 3x$

$I = \frac{1}{3}e^{2x} \sin 3x - \int \frac{2}{3}e^{2x} \sin 3x$

$u = \frac{2}{3}e^{2x}$ $\quad\quad v' = \sin 3x$

$u' = \frac{4}{3}e^{2x}$ $\quad\quad v = \frac{1}{3}\cos 3x$

$I = \frac{1}{3}e^{2x} \sin 3x + \left[\frac{2}{9}e^{2x} \cos 3x + \int \frac{4}{9}e^{2x} \cos 3x\right]$

$\therefore I = \frac{1}{3}e^{2x} \sin 3x - \frac{2}{9}e^{2x} \cos 3x - \frac{4}{9}I + c$

$\therefore \frac{13}{9}I = \frac{1}{3}e^{2x} \sin 3x - \frac{2}{9}e^{2x} \cos 3x + c$

$\quad\quad = \frac{3}{13} \cdot \frac{9}{3}e^{2x} \sin 3x - \frac{9}{13} \cdot \frac{2}{9}e^{2x} \cos 3x + d$

$\quad\quad = \frac{1}{13}e^{2x} [3 \sin 3x + 2 \cos 3x] + d$

DON'T FORGET

Where Integrand is repeated, take extra care with ∴.

ONLINE

Explore this topic further at www.brightredbooks.net

ONLINE TEST

Test yourself on integration at www.brightredbooks.net

 **THINGS TO DO AND THINK ABOUT**

1   Use integration by parts to obtain $\int 10x^2 \cos 5x \, dx$.   **5**

2   Use integration by parts to obtain the exact value of $\int_0^1 x \tan^{-1}x^2 \, dx$.   **5**

3   Use integration by parts to obtain $\int \frac{(x + 3)^2}{(x + 1)^3} \, dx$.   **5**

   Let $I = \int_0^{\pi/2} e^{2x}\cos x \, dx$ and $J = \int_0^{\pi/2} e^{2x}\sin x \, dx$.

   Use integration by parts to show that $I = e^\pi - 2J$, and also that $J = 1 + 2I$.

   Hence show that $\int_0^{\pi/2} e^{2x}\cos x \, dx = \frac{1}{5}(e^\pi - 2)$.   **6**

EQUATIONS

SYSTEMS OF LINEAR EQUATIONS AND GAUSSIAN ELIMINATION

ONLINE

Learn more about systems of equations by following the link at www.brightredbooks.net

OVERVIEW

Linear equations only have variables to the power of 1 and no products of variables. Many problems reduce to solving a system of linear equations; and there is a systematic way of doing this.

Example: 5.1

Consider the following system of equations:

$2x + y + 3z = 6, \quad x + 2y - z = -1, \quad x + y + 2z = 3.$

There are three main steps to solving this system.

Step 1: Write the system of equations as an **augmented matrix**. The numbers (elements) in the first column of this matrix are the coefficients of x, those in the second column the coefficients of y, and so on.

$$\begin{array}{ccc|c} 2 & 1 & 3 & 6 \\ 1 & 2 & -1 & -1 \\ 1 & 1 & 2 & 3 \end{array}$$

We now describe a set of operations, called **elementary row operations**, which replace this matrix by a simpler matrix from which we can easily obtain the solution of our system of linear equations.

The object is to reduce the initial matrix to upper (or lower) triangular form. This means producing zeros in the bottom left-hand triangle for the upper triangular form, or for the lower triangular form in the top right-hand triangle.

Upper triangular form

$$\begin{array}{ccc} 2 & 1 & -3 \\ 0 & 5 & 2 \\ 0 & 0 & 4 \end{array}$$

Lower triangular form

$$\begin{array}{ccc} 2 & 0 & 0 \\ 3 & -1 & 0 \\ -4 & 3 & 1 \end{array}$$

DON'T FORGET

There are three elementary row operations:
- any row of the matrix can be multiplied by any (non-zero) number
- any row can be changed by adding any multiple of another row to it
- any two rows can be interchanged.

Step 2: Simplify the augmented matrix using elementary row operations.

$$\begin{array}{ccc|c} 2 & 1 & 3 & 6 \\ 1 & 2 & -1 & -1 \\ 1 & 1 & 2 & 3 \end{array} \quad \text{Augmented matrix}$$

becomes

$$\begin{array}{ccc|c} 1 & 2 & -1 & -1 \\ 2 & 1 & 3 & 6 \\ 1 & 1 & 2 & 3 \end{array}$$

(interchanging rows 1 and 2)

If possible, start with a 1 in the top left corner of the augmented matrix. Interchanging two rows may do this. It will avoid the use of fractions to begin with.

We now subtract 2 × row 1 from row 2, for which we use the shorthand notation $R'_2 = R_2 - 2R_1$, and also subtract row 1 from row 3, $R'_3 = R_3 - R_1$:

$$\begin{array}{ccc|cl} 1 & 2 & -1 & -1 & \\ 0 & -3 & 5 & 8 & R_2 - 2R_1 \\ 0 & -1 & 3 & 4 & R_3 - R_1 \end{array}$$

DON'T FORGET

Interchange rows if it will give a 1 in the top left corner of the matrix.

The next step is to get a zero in row 3 column 2. We accomplish this by $R'_3 = 3R_3 - R_2$, where R_2 and R_3 now refer to the new version of the original augmented matrix. (Note that there are a number of acceptable alternative shorthand notations.)

$$\begin{array}{ccc|cl} 1 & 2 & -1 & -1 & \\ 0 & -3 & 5 & 8 & \\ 0 & 0 & 4 & 4 & 3R_3 - R_2 \end{array}$$

contd

This last matrix is in **upper triangular form**, the simplified form required.

Step 3: Solve the system of linear equations. The system given initially:

$2x + y + 3z = 6,\quad x + 2y - z = -1,\quad x + y + 2z = 3$

is equivalent to the system obtained from the final matrix:

$x + 2y - z = -1,\quad -3y + 5z = 8,\quad 4z = 4.$

This is because every row operation on the augmented matrix corresponds to an equivalent operation on the system of equations **which does not affect the solution**. This process is familiar to you from previous years' work on simultaneous equations involving only two variables.

Working backwards (back substitution) with the simplified system, we have:

$z = 1$, $-3y + 5 = 8$, so $y = -1$, and $x - 2 - 1 = -1$, so $x = 2$.

 **DON'T FORGET**

1 Augmented matrix
2 Simplify
3 Solve/interpret

The method of reducing a system of equations using elementary row operations, followed by back substitution, is known as **Gaussian elimination**, after the German mathematician Karl Friedrich Gauss (1777–1855), widely regarded as one of the greatest mathematicians ever. Although the technique was known to Chinese mathematicians nearly 2000 years prior to Gauss's time, he established its secure basis within a general theory of equation-solving.

 THINGS TO DO AND THINK ABOUT

1 Use Gaussian elimination to obtain solutions of the equations:
$x + 3y + 5z = 14$
$2x - y - 3z = 3$
$4x + 5y - z = 7.$ 5

2 Use Gaussian elimination to solve the following system of equations:
$x + y - z = 6$
$2x - 3y + 2z = 2$
$-5x + 2y - 4z = 1.$ 5

ONLINE TEST

Test yourself on equations at www.brightredbooks.net

SPECIAL CASES

REDUNDANCY

Redundancy means that one of the equations can be derived from the others, so adds no further information. It arises when a plane, given by one of the equations, contains the line of intersection of two other planes, given by the other two equations.

 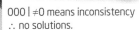
Example: 5.2

Solve the system of equations $x - 2y + z = 1$, $3x + y - z = 0$, $5x - 3y + z = 2$.

Solution:

$$\begin{array}{ccc|c} 1 & -2 & 1 & 1 \\ 3 & 1 & -1 & 0 \\ 5 & -3 & 1 & 2 \end{array} \qquad \begin{array}{ccc|cl} 1 & -2 & 1 & 1 \\ 0 & 7 & -4 & -3 & R_2 - 3R_1 \\ 0 & 7 & -4 & -3 & R_3 - 5R_1 \end{array}$$

Note that the second and third rows are the same, which leads to:

$$\begin{array}{ccc|cl} 1 & -2 & 1 & 1 \\ 0 & 7 & -4 & -3 \\ 0 & 0 & 0 & 0 & R_3 - R_2 \end{array}$$

Because there is no equation for z, **z can have any value**. Set $z = t$ to emphasise this, where t is a parameter which can take any value in $\mathbb{R}$.

Solving for y then x gives $z = t$, $y = -\frac{3}{7} + \frac{4}{7}t$, $x = \frac{1}{7} + \frac{t}{7}$

These equations give points that all lie on a straight line (see pp 68–81).

The system of equations has an infinite number of solutions. There are, really, only two equations, and the third is redundant.

INCONSISTENCY

Inconsistency means that there are no values of x, y, z which satisfy all three equations. It arises when three planes, given by the three equations, do not have a point in common.

 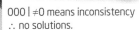
Example: 5.3

Solve the system of equations $x - y - 2z = 0$, $3x + 2y - z = 2$, $2x + 3y + z = 1$.

Solution:

$$\begin{array}{ccc|c} 1 & -1 & -2 & 0 \\ 3 & 2 & -1 & 2 \\ 2 & 3 & 1 & 1 \end{array}$$

$$\begin{array}{ccc|cl} 1 & -1 & -2 & 0 \\ 0 & 5 & 5 & 2 & R_2 - 3R_1 \\ 0 & 5 & 5 & 1 & R_3 - 2R_1 \end{array}$$

$$\begin{array}{ccc|cl} 1 & -1 & -2 & 0 \\ 0 & 5 & 5 & 2 \\ 0 & 0 & 0 & 1 & R_2 - R_3 \end{array}$$

This last third row says that $0x + 0y + 0z = 1$, i.e. $0 = 1$, which is clearly impossible. The system of equations has no solution, and is said to be inconsistent.

Geometrical interpretations of redundancy and inconsistency in terms of the intersection of three planes are given in Vectors: pp 68–81.

ILL-CONDITIONING

All our calculations so far have been exact. However, in many real-life problems using experimental or collected data, the coefficients in the equations are often rounded to the nearest integer, or one decimal place, and so on. We will look at a couple of examples where some of the coefficients have been rounded to the nearest integer.

Example: 5.4

The equations

$$2x + y = 1 \quad \text{and} \quad 2x - y = 2$$

have the solution $x = 0{\cdot}75$, $y = -0{\cdot}5$.

Suppose that the coefficients of x have been rounded and that the equations should be:

$$1{\cdot}9x + y = 1$$
$$1{\cdot}9x - y = 2.$$

The solution is now $x = 0{\cdot}8$, $y = -0{\cdot}5$, to 1 decimal place.

So, a 5% change in some of the coefficients has led to about a 6% change in the answer. A situation like this in a real-life problem would be satisfactory.

ONLINE

Learn about redundancy and inconsistency at www.brightredbooks.net

Example: 5.5

The equations

$$2x - y = -5 \ (20x - 10y = -50) \quad \text{and} \quad 21x - 10y = -10$$

have the solution $x = 40$ and $y = 85$.

Now suppose that the equations should be:

$$19x - 10y = -50 \quad \text{and} \quad 22x - 10y = -10.$$

The solution now is $x = 13\frac{1}{3}$ and $y = 30\frac{1}{3}$.

This time, a 5% change in some of the coefficients has led to a 67% change in x and a 64% change in y. This set of equations is said to be **ill-conditioned**. A small percentage change in the coefficients has led to a significantly larger percentage change in the solution.

This set of equations would not therefore form a good model for real-life contexts.

ONLINE

If you want to read further about ill-conditioning, go to www.brightredbooks.net

Both of these examples involve calculating the point of intersection of a pair of straight lines. In example 5.4, the angle between the lines is large (their gradients are 2 and −2). However, in example 5.5, the two lines are nearly parallel (their gradients are 2 and 2·1). A small change in any of the coefficients in the equations represents a small change in either line; and, because they are nearly parallel, this produces a big change in the point of intersection. Similarly, where the gradients of the two lines representing the equations is similar, a small change in the constant may produce a disproportionate change in the solution, again resulting in **ill-conditioning**.

A similar situation holds for some systems of linear equations in three unknowns, where now the solution represents the intersection of planes.

ONLINE TEST

Test yourself on equations at www.brightredbooks.net

 ## THINGS TO DO AND THINK ABOUT

1 Use Gaussian elimination to solve the following system of equations:

   $$x + 3y + 2z = 4 \qquad 2x + 4y - 5z = -8 \qquad x - 3y - 25z = -44 \qquad \qquad 5$$

2 Use Gaussian elimination to solve the system of equations below when $\lambda \neq 6$.
   Explain what happens when $\lambda = 6$.   **5, 1**

   $$x + y + 2z = 3 \qquad 2x + 3y + 4z = 5 \qquad 3x + 2y + \lambda z = 11$$

3 **(a)** Show that the following set of equations can be solved to produce a unique solution when $\lambda \neq 7$.

   $$x - y + z = -7 \quad x + y + 3z = -1 \qquad 3x + y + \lambda z = -9$$

   State the solution.   **5**

   (b) Explain what happens when $\lambda = 7$.   **1**

   (c) Give the solutions to (b) in terms of z.   **3**

MATRICES

TERMS AND BASIC MATRIX OPERATIONS

OVERVIEW

ONLINE

Follow the link at www.brightredbooks.net to find out more about matrices.

Example: 6.1

$$A = \begin{bmatrix} 1 & 3 \\ 2 & 1 \end{bmatrix}, B = \begin{bmatrix} 5 & 1 & 3 \\ -1 & 0 & x \end{bmatrix}, C = \begin{bmatrix} 2 \\ 1 \\ 1 \end{bmatrix} \text{ are matrices.}$$

A is a 2 × 2 matrix (read as *two by two*), and is also a **square matrix**. It has **order** 2 × 2.

B has order 2 × 3, having two **rows** and three **columns**.

C is a **column matrix**.

DON'T FORGET

Order: row by column.

Matrices are particularly useful when we define algebraic operations on them.

Two matrices are said to be **equal** when they have the same order and all corresponding elements are equal.

Example: 6.2

The matrices $\begin{bmatrix} 1 & 3 \\ 2 & 1 \end{bmatrix}$ and $\begin{bmatrix} 3 & 1 \\ 1 & 2 \end{bmatrix}$ are not equal.

$\begin{bmatrix} 1 & 3 \\ 2 & 1 \end{bmatrix}$ and $\begin{bmatrix} 1 & 3 \\ x & 1 \end{bmatrix}$ are only equal when $x = 2$.

ADDITION, SUBTRACTION AND MULTIPLYING BY A SCALAR

Only matrices of the same order can be added. Simply add corresponding elements.

DON'T FORGET

Matrix addition looks like ordinary algebraic addition.
- $A + B = B + A$,
- $(A + B) + C$
  $= A + (B + C)$.

Example: 6.3

$$\begin{bmatrix} 1 & 3 \\ 2 & 1 \end{bmatrix} + \begin{bmatrix} 2 & -1 \\ 0 & 3 \end{bmatrix} = \begin{bmatrix} 3 & 2 \\ 2 & 4 \end{bmatrix} \text{ and } \begin{bmatrix} 5 & 1 & 3 \\ -1 & 0 & x \end{bmatrix} + \begin{bmatrix} 2 & 1 & -3 \\ 3 & 2x & x \end{bmatrix} = \begin{bmatrix} 7 & 2 & 0 \\ 2 & 2x & 2x \end{bmatrix}$$

In example 6.1 above, the matrices A and B, A and C, B and C cannot be added.

We naturally denote $A + A$ by $2A$, so we define rA for any number r to be the matrix obtained from A by multiplying all the elements of A by r.

Example: 6.4

$$2\begin{bmatrix} 1 & 3 \\ -4 & 0 \\ 5 & -1 \end{bmatrix} = \begin{bmatrix} 2 & 6 \\ -8 & 0 \\ 10 & -2 \end{bmatrix}$$

Example: 6.5

$$2\begin{bmatrix} 1 & 3 \\ 2 & 1 \end{bmatrix} - 3\begin{bmatrix} 2 & -1 \\ 0 & 3 \end{bmatrix} = \begin{bmatrix} 2 & 6 \\ 4 & 2 \end{bmatrix} - \begin{bmatrix} 6 & -3 \\ 0 & 9 \end{bmatrix} = \begin{bmatrix} -4 & 9 \\ 4 & -7 \end{bmatrix}$$

MATRIX MULTIPLICATION

Multiplying a row matrix of order $1 \times n$ with a column matrix of order $n \times 1$, the product is given by:

$$n = 2: \begin{bmatrix} a & b \end{bmatrix} \begin{bmatrix} c \\ d \end{bmatrix} = ac + bd; \quad n = 3: \begin{bmatrix} a & b & c \end{bmatrix} \begin{bmatrix} d \\ e \\ f \end{bmatrix} = ad + be + cf; \text{ and so on.}$$

In each case, the result is a number which is thought of as a 1×1 matrix.

In general, to define the product AB of two matrices, in the order written, the number of columns in A must equal the number of rows in B. Treat each row of A as a row matrix, and each column of B as a column matrix, and proceed as follows:

- Multiply row 1 of A by column 1 of B and put the result in row 1 column 1.

- Multiply row 1 of A by column 2 of B and put the result in row 1 column 2.

- When the columns of B have all been used, move to row 2 of A.

- Multiply row 2 of A by column 1 of B and put the result in row 2 column 1.

- Continue until every row of A has been multiplied by every column of B.

The diagram on the right shows how to multiply a 2×3 matrix on the left by a 3×2 matrix on the right to produce a 2×2 matrix:

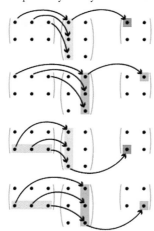

Example: 6.6

Let $A = \begin{bmatrix} 1 & 2 \\ 0 & 1 \end{bmatrix}$, $B = \begin{bmatrix} 2 & 1 \\ 4 & 3 \end{bmatrix}$

Then $AB = \begin{bmatrix} 1 \times 2 + 2 \times 4 & 1 \times 1 + 2 \times 3 \\ 0 \times 2 + 1 \times 4 & 0 \times 1 + 1 \times 3 \end{bmatrix} = \begin{bmatrix} 10 & 7 \\ 4 & 3 \end{bmatrix}$

$BA = \begin{bmatrix} 2 \times 1 + 1 \times 0 & 2 \times 2 + 1 \times 1 \\ 4 \times 1 + 3 \times 0 & 4 \times 2 + 3 \times 1 \end{bmatrix} = \begin{bmatrix} 2 & 5 \\ 4 & 11 \end{bmatrix}$

Note that $AB \neq BA$.

 THINGS TO DO AND THINK ABOUT

1 Find AB and BA when $A = \begin{bmatrix} 2 & -3 \\ 1 & 4 \end{bmatrix}$ and $B = \begin{bmatrix} -1 & 5 \\ 0 & -2 \end{bmatrix}$. 　2

2 Find CD when $C = \begin{bmatrix} 2 & 3 & 4 \\ 1 & 0 & -1 \\ 3 & -2 & -1 \end{bmatrix}$ and $D = \begin{bmatrix} 6 & 0 \\ -1 & 2 \\ 4 & -3 \end{bmatrix}$. 　2

 ONLINE TEST

Test yourself on matrices at www.brightredbooks.net

MATRIX MULTIPLICATION AND THE TRANSPOSE OF A MATRIX GPS

MATRIX MULTIPLICATION (CONTD.)

Example: 6.7

$$\begin{bmatrix} 2 & 0 & 1 \\ -2 & 3 & 1 \\ 1 & -1 & 0 \end{bmatrix} \begin{bmatrix} 2 \\ 1 \\ 2 \end{bmatrix} = \begin{bmatrix} 2 \times 2 + 0 \times 1 + 1 \times 2 \\ (-2) \times 2 + 3 \times 1 + 1 \times 2 \\ 1 \times 2 + (-1) \times 1 + 0 \times 2 \end{bmatrix} = \begin{bmatrix} 6 \\ 1 \\ 1 \end{bmatrix}$$

DON'T FORGET

Usually, *AB* is not the same as *BA*.

Note: when calculating AB, all answers to multiplying by row 1 of A go in row 1 in the resultant matrix, all for row 2 of A go in row 2, and so on. Similarly, all answers to multiplying by column 1 of B go in column 1 of the resultant matrix, and so on.

Although $AB \neq BA$ in general, for some matrices $AB = BA$.

Example: 6.8

$$\begin{bmatrix} 1 & -1 \\ 2 & 0 \end{bmatrix} \begin{bmatrix} 2 & 1 \\ -2 & 3 \end{bmatrix} = \begin{bmatrix} 4 & -2 \\ 4 & 2 \end{bmatrix} = \begin{bmatrix} 2 & 1 \\ -2 & 3 \end{bmatrix} \begin{bmatrix} 1 & -1 \\ 2 & 0 \end{bmatrix}$$

Matrix algebra looks very much like ordinary algebra.

In general, $AB \neq BA$, but

$$(AB)C = A(BC)$$
$$A(B + C) = AB + AC$$

and $\qquad AA = A^2$.

THE TRANSPOSE OF A MATRIX

If we interchange the rows and columns of a matrix A, the resulting matrix is denoted by A', or A^T, and called the **transpose** of A.

Example: 6.9

If $A = \begin{bmatrix} 1 & 0 & 2 \\ 2 & 1 & 3 \end{bmatrix}$, $B = \begin{bmatrix} 2 & 0 \\ -1 & 5 \end{bmatrix}$, $C = \begin{bmatrix} 1 & 3 & 0 \end{bmatrix}$,

then $A' = \begin{bmatrix} 1 & 2 \\ 0 & 1 \\ 2 & 3 \end{bmatrix}$, $B' = \begin{bmatrix} 2 & -1 \\ 0 & 5 \end{bmatrix}$, $C' = \begin{bmatrix} 1 \\ 3 \\ 0 \end{bmatrix}$.

Points to note: $(A')' = A$

$(A + B)' = A' + B'$

$(AB)' = B'A'$. Note the order on the right-hand side.

Check this for the matrices A, B in example 6.6.

contd

If $A = A'$ for a matrix A, we say that A is symmetric.

Example: 6.10

$$\begin{bmatrix} 2 & 3 & 5 \\ 3 & 0 & 6 \\ 5 & 6 & -1 \end{bmatrix}$$ is symmetric.

We can reflect in the leading diagonal: $\begin{bmatrix} 2 & & \\ & 0 & \\ & & -1 \end{bmatrix}$

Let's think about this

If a matrix is symmetric, why must it be a square matrix?

Answer: If A has order $m \times n$, then A' has order $n \times m$, so they can only be equal when $m = n$.

 THINGS TO DO AND THINK ABOUT

1   Matrices A, B and C are given by:
$A = \begin{bmatrix} 2 & 5 \\ -3 & 7 \end{bmatrix}$; $B = \begin{bmatrix} 1 & -2b \\ b & 1 \end{bmatrix}$; $C = \begin{bmatrix} 5 & q \\ p & 17 \end{bmatrix}$.

(a) Obtain a single matrix for $B^{T}A$. **2**

(b) Given that $B^{T}A = C$, find the values of b, p and q. **3**

2   Two 3×3 matrices are given by:
$A = \begin{bmatrix} 2 & 3 & 1 \\ p & -1 & 0 \\ 2 & 1 & 1 \end{bmatrix}$; $B = \begin{bmatrix} 1 & 1 & 2 \\ 2 & -1 & 0 \\ 0 & 2 & -2 \end{bmatrix}$.

(a) Express $3A - 5B$ as a single matrix. **2**

(b) Express AB as a single matrix. **2**

(c) Given that AB is singular, calculate p. **3**

ONLINE

Learn more about multiplying matrices by following the link at www.brightredbooks.net

ONLINE

Networks and matrices: businesses and governments are increasingly using networks as a planning tool. They usually wish to find the optimum solution to a practical problem. Discover more at www.brightredbooks.net

ONLINE TEST

Test yourself on matrices at www.brightredbooks.net

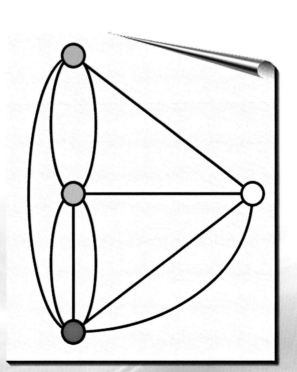

SPECIAL MATRICES 1 GPS

IDENTITY MATRIX

For 2 × 2 matrices, the matrix $I = \begin{bmatrix} 1 & 0 \\ 0 & 1 \end{bmatrix}$ satisfies $AI = IA = A$ for **all** 2 × 2 matrices A.

I is called the **identity** (or **unit**) matrix.

For 3 × 3 matrices, the matrix $I = \begin{bmatrix} 1 & 0 & 0 \\ 0 & 1 & 0 \\ 0 & 0 & 1 \end{bmatrix}$ satisfies $AI = IA = A$ for **all** 3 × 3 matrices A.

Note that we use the letter I for both the 2 × 2 case and the 3 × 3 case. The context makes it clear which one we mean.

DETERMINANTS AND INVERSES OF SQUARE MATRICES

For a given square matrix A, if we can find a matrix B such that $AB = I$, B is called the **inverse** of A, and is denoted by A^{-1}. When A^{-1} exists, we have $AA^{-1} = I = A^{-1}A$.

Inverses of 2 × 2 matrices

The expression $D = ad - bc$ is called the **determinant** of the matrix $\begin{bmatrix} a & b \\ c & d \end{bmatrix}$.

The determinant of a matrix A is often denoted by **det A** or by $\begin{vmatrix} a & b \\ c & d \end{vmatrix}$.

The inverse of A is given by $A^{-1} = \frac{1}{D} \begin{bmatrix} d & -b \\ -c & a \end{bmatrix}$.

If $D = 0$, then the matrix A doesn't have an inverse; A is said to be **singular**.

Example: 6.11

Let $A = \begin{bmatrix} 2 & 1 \\ -1 & 3 \end{bmatrix}$ and $B = \begin{bmatrix} 2 & 1 \\ 4 & 2 \end{bmatrix}$.

Then $\det A = 2 \times 3 - (-1) \times 1 = 7$ and $\det B = 2 \times 2 - 4 \times 1 = 0$.

So, A is **non-singular**, and $A^{-1} = \frac{1}{7} \begin{bmatrix} 3 & -1 \\ 1 & 2 \end{bmatrix} = \begin{bmatrix} \frac{3}{7} & -\frac{1}{7} \\ \frac{1}{7} & \frac{2}{7} \end{bmatrix}$.

B is singular; it doesn't have an inverse.

Determinant of 3 × 3 matrices

The determinant of a 3 × 3 matrix determines when the matrix has an inverse, just as in the 2 × 2 case.

If $A = \begin{bmatrix} a_1 & b_1 & c_1 \\ a_2 & b_2 & c_2 \\ a_3 & b_3 & c_3 \end{bmatrix}$, then $\det A = a_1 \begin{vmatrix} b_2 & c_2 \\ b_3 & c_3 \end{vmatrix} - b_1 \begin{vmatrix} a_2 & c_2 \\ a_3 & c_3 \end{vmatrix} + c_1 \begin{vmatrix} a_2 & b_2 \\ a_3 & b_3 \end{vmatrix}$.

So, $\det A = a_1(b_2 c_3 - b_3 c_2) - b_1(a_2 c_3 - a_3 c_2) + c_1(a_2 b_3 - a_3 b_2)$.

Note the negative sign before the term in b_1.

Example: 6.12

$\text{Det} \begin{bmatrix} 2 & 1 & 2 \\ 1 & -1 & 0 \\ 3 & 2 & 2 \end{bmatrix} = 2 \begin{vmatrix} -1 & 0 \\ 2 & 2 \end{vmatrix} - 1 \begin{vmatrix} 1 & 0 \\ 3 & 2 \end{vmatrix} + 2 \begin{vmatrix} 1 & -1 \\ 3 & 2 \end{vmatrix} = 2(-2 - 0) - (2 - 0) + 2(2 - (-3)) = 4.$

contd

Alternative method for computing the determinant

In the 1800s, Pierre Frédéric Sarrus discovered a mnemonic rule for solving the determinant of a 3 × 3 matrix, named Sarrus's scheme.

Sarrus's rule or **Sarrus's scheme** is a method and a memorisation scheme to compute the determinant of a 3 × 3 matrix.

Consider a 3 × 3 matrix.

$$\begin{bmatrix} 1 & 2 & 3 \\ 4 & 5 & 6 \\ 7 & 8 & 9 \end{bmatrix}$$

ONLINE

Learn more about Pierre Frédéric Sarrus by following the link at www.brightredbooks.net

Then its determinant can be computed by writing the first two columns of the matrix to the right of the third column so that there are five columns.

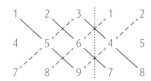

The determinant of the three columns on the left is the sum of the products along the solid diagonals minus the sum of the products along the dashed diagonals.

Here we have

$[(1 \times 5 \times 9) + (2 \times 6 \times 7) + (3 \times 4 \times 8)]$

$- [(7 \times 5 \times 3) + (8 \times 6 \times 1) + (9 \times 4 \times 2)]$

$= (45 + 84 + 96) - (105 + 48 + 72) = 225 - 225 = 0$

VIDEO LINK

Watch the clips at www.brightredbooks.net to learn more about this topic.

You may find it easier to use the alternative vertical arrangement where the first and second rows are written under the third row.

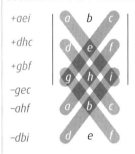

ONLINE TEST

Test yourself on special matrices at www.brightredbooks.net

 THINGS TO DO AND THINK ABOUT

1 Given $A = \begin{bmatrix} 2 & -3 \\ 1 & 4 \end{bmatrix}$ and $B = \begin{bmatrix} -1 & 5 \\ 0 & -2 \end{bmatrix}$, find the determinants of A and B.　2

2 Use the first method above, and then try Sarrus's rule for finding the determinant of matrix C, where

$C = \begin{bmatrix} 2 & 3 & 4 \\ 1 & 0 & -1 \\ 3 & -2 & -1 \end{bmatrix}$.　2

3 Given the matrix $A = \begin{bmatrix} t+4 & 3t \\ 3 & 5 \end{bmatrix}$,

   (a) Find A^{-1} in terms of t when A is non-singular.　3

   (b) Write down the value of t such that A is singular.　1

   (c) Given that the transpose of A is $\begin{bmatrix} 6 & 3 \\ 6 & 5 \end{bmatrix}$, find t.　1

SPECIAL MATRICES 2 GPS

INVERSES OF 3 × 3 MATRICES

Just as for 2 × 2 matrices (in fact, for all square matrices), a 3 × 3 matrix has an inverse if, and only if, its determinant is non-zero.

To calculate the inverse of a 3 × 3 matrix (when it has an inverse), we use elementary row operations as follows.

Start with the augmented matrix $A|I$ and reduce it to the form $I|B$ by elementary row operations. Then $B = A^{-1}$.

In $A|I$, start by reducing A to A_1, which is in upper triangular form (I changes to I_1 when you do this), then reduce A_1 to lower triangular form A_2 (I_1 changes to I_2). Now reduce A_2 to I, changing to B, the inverse of A.

There are lots of ways in which elementary row operations can be used to get to the final form, so the steps in example 6.13 below are not the only ones. There are also other valid methods for finding the inverse.

Example: 6.13

Calculate the inverse of $\begin{bmatrix} 2 & 1 & 2 \\ 1 & -1 & 0 \\ 3 & 2 & 2 \end{bmatrix}$.

(We worked horizontally here to save space, but your working should progress **down** the page.)

$$\begin{array}{ccc|ccc}
2 & 1 & 2 & 1 & 0 & 0 \\
1 & -1 & 0 & 0 & 1 & 0 \\
3 & 2 & 2 & 0 & 0 & 1
\end{array}
\longrightarrow
\begin{array}{ccc|ccc}
2 & 1 & 2 & 1 & 0 & 0 \\
0 & -3 & -2 & -1 & 2 & 0 \\
0 & 1 & -2 & -3 & 0 & 2
\end{array}
\begin{array}{l} \\ 2R_2 - R_1 \\ 2R_3 - 3R_1 \end{array}$$

$$\begin{array}{ccc|ccc}
2 & 1 & 2 & 1 & 0 & 0 \\
0 & -3 & -2 & -1 & 2 & 0 \\
0 & 0 & -8 & -10 & 2 & 6
\end{array}
\begin{array}{l} \\ \\ 3R_3 + R_2 \end{array}
\longrightarrow
\begin{array}{ccc|ccc}
8 & 4 & 0 & -6 & 2 & 6 \\
0 & -12 & 0 & 6 & 6 & -6 \\
0 & 0 & -8 & -10 & 2 & 6
\end{array}
\begin{array}{l} 4R_1 + R_3 \\ 4R_2 - R_3 \\ \end{array}$$

$$\begin{array}{ccc|ccc}
24 & 0 & 0 & -12 & 12 & 12 \\
0 & -12 & 0 & 6 & 6 & -6 \\
0 & 0 & -8 & -10 & 2 & 6
\end{array}
\begin{array}{l} 3R_1 + R_2 \\ \\ \end{array}$$

Now we just divide row 1 by 24, row 2 by −12, and row 3 by −8 to give:

$$\begin{array}{ccc|ccc}
1 & 0 & 0 & -\frac{1}{2} & \frac{1}{2} & \frac{1}{2} \\
0 & 1 & 0 & -\frac{1}{2} & -\frac{1}{2} & \frac{1}{2} \\
0 & 0 & 1 & \frac{5}{4} & -\frac{1}{4} & -\frac{3}{4}
\end{array}$$

So, the inverse of $\begin{bmatrix} 2 & 1 & 2 \\ 1 & -1 & 0 \\ 3 & 2 & 2 \end{bmatrix}$ is $\begin{bmatrix} -\frac{1}{2} & \frac{1}{2} & \frac{1}{2} \\ -\frac{1}{2} & -\frac{1}{2} & \frac{1}{2} \\ \frac{5}{4} & -\frac{1}{4} & -\frac{3}{4} \end{bmatrix}$.

If you reduce $A|I$ to $E|F$, where

$$E = \begin{bmatrix} p & 0 & 0 \\ 0 & q & 0 \\ 0 & 0 & r \end{bmatrix}$$, and p, q, r are integers, you can avoid fractions until the last step.

As it requires so much arithmetic to obtain the inverse of a 3 × 3 matrix, many examination questions do not require you to use elementary row operations. The next example is typical.

contd

Example: 6.14

Let $A = \begin{bmatrix} 1 & 2 & 1 \\ 2 & -1 & 1 \\ 1 & 0 & 2 \end{bmatrix}$ and $B = \begin{bmatrix} 2 & 4 & -3 \\ 3 & x & -1 \\ x & -2 & 5 \end{bmatrix}$. Calculate AB and hence, or otherwise, obtain A^{-1}.

$AB = \begin{bmatrix} 8+x & 2+2x & 0 \\ x+1 & 6-x & 0 \\ 2+2x & 0 & 7 \end{bmatrix}$

If we choose $x = -1$, we get $AB = \begin{bmatrix} 7 & 0 & 0 \\ 0 & 7 & 0 \\ 0 & 0 & 7 \end{bmatrix} = 7I$ and $B = \begin{bmatrix} 2 & 4 & -3 \\ 3 & -1 & -1 \\ -1 & -2 & 5 \end{bmatrix}$. Hence $A^{-1} = \frac{1}{7}B = \begin{bmatrix} \frac{2}{7} & \frac{4}{7} & -\frac{3}{7} \\ \frac{3}{7} & -\frac{1}{7} & -\frac{1}{7} \\ -\frac{1}{7} & -\frac{2}{7} & \frac{5}{7} \end{bmatrix}$.

In questions like this, there are usually only 1 or 2 marks available for obtaining A^{-1}, so although you **could** use elementary row operations to obtain it, you should be on the lookout for the quicker method.

Example: 6.15

Given two matrices A and B of the same order (any order), and given A^{-1}, B^{-1}, how do we get the inverses of AB and BA?

We have $(AB)(B^{-1}A^{-1}) = A(BB^{-1})A^{-1} = AIA^{-1} = AA^{-1} = I$, because brackets can go anywhere in a product (but don't change the order of the product).

This shows that the inverse of AB is $B^{-1}A^{-1}$ (note the order). In the same way, the inverse of BA is $A^{-1}B^{-1}$.

 **DON'T FORGET**

$(AB)^{-1} = B^{-1}A^{-1}$

Let's think about this

Why can't we define $\frac{A}{B}$ for matrices A and B, even when B is non-singular?

Answer: We could (but we don't) define $\frac{1}{B}$ to be B^{-1}, but then we wouldn't know whether $\frac{A}{B}$ meant AB^{-1} or $B^{-1}A$.

 **DON'T FORGET**

You cannot divide matrices.

 VIDEO LINK

Here EROs (elementary row operations) have been used to find the inverse (and you can see this method in a video at www.brightredbooks.net).

 THINGS TO DO AND THINK ABOUT

1   $A = \begin{bmatrix} 2 & 1 \\ a & 0 \end{bmatrix}$

    (a) Find A^2 as a single matrix.   1
    (b) Given that $A^2 = 2A + 3I$, find the value of a.   3
    (c) Express A^3 in the form $kA + mI$.   3

2   Calculate the inverse of the matrix $\begin{bmatrix} 1 & 2 \\ -x & 3 \end{bmatrix}$. For what value of x is this matrix singular? 4

 ONLINE

There are other methods – find out about the adjoint and inverse of a matrix; and how the inverse of a matrix can be found by using minors, cofactors and adjugate at www.brightredbooks.net

3   $A = \begin{bmatrix} 6 & -2 & 2 \\ -3 & 1 & -3 \\ 1 & 1 & 1 \end{bmatrix}$ $B = \begin{bmatrix} 1 & 1 & 1 \\ 0 & 1 & 3 \\ -1 & -2 & 0 \end{bmatrix}$

    (a) Calculate AB.   1
    (b) Hence state matrix B in terms of matrix A.   2
    (c) Given $C = BA$, write down matrix C.   1

4   Given the matrix $A = \begin{bmatrix} \lambda & 2 \\ \lambda+3 & 4 \end{bmatrix}$,

    (a) Obtain A^{-1} when A is non-singular.   3
    (b) For what value of λ is A singular?   1
    (c) Given that $A' = \begin{bmatrix} -2 & 1 \\ 2 & 4 \end{bmatrix}$, obtain the value of λ.   1

 ONLINE TEST

Test yourself on matrices at www.brightredbooks.net

5   Find the value of k such that the matrix $A = \begin{bmatrix} k & 1 & -1 \\ -2 & -3 & 0 \\ k-1 & 2 & 1 \end{bmatrix}$ is singular.   3

6   A square matrix is such that:
    $A^3 = 13A - 12I$ and $A^4 = 40A - 39I$.
    (a) Show that A^{-1} can be expressed in the form $-\frac{1}{3}A + \frac{4}{3}I$.   4
    (b) Hence express A^2 in the form $pA + qI$, stating the values of p and q.   4

LINEAR EQUATIONS IN MATRIX FORM AND GEOMETRIC TRANSFORMATIONS GPS

LINEAR EQUATIONS IN MATRIX FORM

The system of linear equations

$$a_1x + b_1y = c_1$$
$$a_2x + b_2y = c_2$$

is equivalent to the matrix equation: $\begin{bmatrix} a_1 & b_1 \\ a_2 & b_2 \end{bmatrix} \begin{bmatrix} x \\ y \end{bmatrix} = \begin{bmatrix} c_1 \\ c_2 \end{bmatrix}$,

that is, $A \begin{bmatrix} x \\ y \end{bmatrix} = \begin{bmatrix} c_1 \\ c_2 \end{bmatrix}$ where $A = \begin{bmatrix} a_1 & b_1 \\ a_2 & b_2 \end{bmatrix}$.

So, when A is non-singular, multiplying on the left by A^{-1} gives

$$A^{-1}\left(A \begin{bmatrix} x \\ y \end{bmatrix} \right) = A^{-1} \begin{bmatrix} c_1 \\ c_2 \end{bmatrix}.$$

But $A^{-1}\left(A \begin{bmatrix} x \\ y \end{bmatrix} \right) = (A^{-1}A) \begin{bmatrix} x \\ y \end{bmatrix}$.

So, $A^{-1}\left(A \begin{bmatrix} x \\ y \end{bmatrix} \right) = (A^{-1}A) \begin{bmatrix} x \\ y \end{bmatrix} = I \begin{bmatrix} x \\ y \end{bmatrix} = \begin{bmatrix} x \\ y \end{bmatrix}$,

hence $\begin{bmatrix} x \\ y \end{bmatrix} = A^{-1} \begin{bmatrix} c_1 \\ c_2 \end{bmatrix}$, from which we get x and y.

So, if we can find A^{-1}, the system of linear equations can be solved.

This works for the 3 × 3 case also.

Example: 6.16

Given that the inverse of $A = \begin{bmatrix} 1 & 2 & 1 \\ 2 & -1 & 1 \\ 1 & 0 & 2 \end{bmatrix}$ is $\begin{bmatrix} \frac{2}{7} & \frac{4}{7} & -\frac{3}{7} \\ \frac{3}{7} & -\frac{1}{7} & -\frac{1}{7} \\ -\frac{1}{7} & -\frac{2}{7} & \frac{5}{7} \end{bmatrix}$,

solve the system of equations

$x + 2y + z = 14, \quad 2x - y + z = 7, \quad x + 2z = -21.$

Solution:

The system of equations may be written in matrix form

$$\begin{bmatrix} 1 & 2 & 1 \\ 2 & -1 & 1 \\ 1 & 0 & 2 \end{bmatrix} \begin{bmatrix} x \\ y \\ z \end{bmatrix} = \begin{bmatrix} 14 \\ 7 \\ -21 \end{bmatrix}$$

which gives:

$$\begin{bmatrix} x \\ y \\ z \end{bmatrix} = A^{-1} \begin{bmatrix} 14 \\ 7 \\ -21 \end{bmatrix} = \begin{bmatrix} \frac{2}{7} & \frac{4}{7} & -\frac{3}{7} \\ \frac{3}{7} & -\frac{1}{7} & -\frac{1}{7} \\ -\frac{1}{7} & -\frac{2}{7} & \frac{5}{7} \end{bmatrix} \begin{bmatrix} 14 \\ 7 \\ -21 \end{bmatrix}$$

$$= \begin{bmatrix} 4 + 4 + 9 \\ 6 - 1 + 3 \\ -2 - 2 - 15 \end{bmatrix} = \begin{bmatrix} 17 \\ 8 \\ -19 \end{bmatrix}$$

Hence $x = 17$, $y = 8$, $z = -19$.

GEOMETRIC TRANSFORMATIONS

Transformations of the plane, such as rotations and reflections, can be represented by matrices. If the transformation maps the points (1, 0) to (a, c) and (0, 1) to (b, d),

it is represented by the matrix $\begin{bmatrix} a & b \\ c & d \end{bmatrix}$.

The point (1, 0) is represented by the matrix $\begin{bmatrix} 1 \\ 0 \end{bmatrix}$

and the point (0, 1) by the matrix $\begin{bmatrix} 0 \\ 1 \end{bmatrix}$.

$$\begin{bmatrix} 1 & 0 \\ 0 & 1 \end{bmatrix} \text{ goes to } \begin{bmatrix} a & b \\ c & d \end{bmatrix}$$
$$\uparrow \quad \uparrow$$
$$(1, 0) \ (0, 1)$$

$$\begin{bmatrix} a & b \\ c & d \end{bmatrix}\begin{bmatrix} x \\ y \end{bmatrix} = \begin{bmatrix} ax + by \\ cx + dy \end{bmatrix}$$

Hence the point $(x, y) \rightarrow (ax + by, cx + dy)$ under this transformation.

$$\begin{bmatrix} -1 & 0 \\ 0 & 1 \end{bmatrix}\begin{bmatrix} x \\ y \end{bmatrix} = \begin{bmatrix} -x \\ y \end{bmatrix}$$

So $(x, y) \rightarrow (-x, y)$, i.e. reflection in the y-axis.

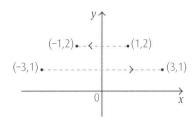

If the matrices A, B represent transformations T_1, T_2 of the plane, then BA represents the transformation given by T_1 followed by T_2.

See table of transformation matrices on p8.

🔵 DON'T FORGET

Note the order of the matrices!

🔵 DON'T FORGET

BA means A first.

Example: 6.17

Write down the matrices A, representing a reflection in the y-axis, and B, representing an anti-clockwise rotation about the origin through 45°. Hence obtain the matrices representing the result of applying:

(a) the reflection followed by the rotation

(b) the rotation followed by the reflection.

Solution:

$$A = \begin{bmatrix} -1 & 0 \\ 0 & 1 \end{bmatrix}, \ B = \begin{bmatrix} \frac{1}{\sqrt{2}} & -\frac{1}{\sqrt{2}} \\ \frac{1}{\sqrt{2}} & \frac{1}{\sqrt{2}} \end{bmatrix}$$

(a) $\begin{bmatrix} \frac{1}{\sqrt{2}} & -\frac{1}{\sqrt{2}} \\ \frac{1}{\sqrt{2}} & \frac{1}{\sqrt{2}} \end{bmatrix}\begin{bmatrix} -1 & 0 \\ 0 & 1 \end{bmatrix} = \begin{bmatrix} -\frac{1}{\sqrt{2}} & -\frac{1}{\sqrt{2}} \\ -\frac{1}{\sqrt{2}} & \frac{1}{\sqrt{2}} \end{bmatrix}$

(b) $\begin{bmatrix} -1 & 0 \\ 0 & 1 \end{bmatrix}\begin{bmatrix} \frac{1}{\sqrt{2}} & -\frac{1}{\sqrt{2}} \\ \frac{1}{\sqrt{2}} & \frac{1}{\sqrt{2}} \end{bmatrix} = \begin{bmatrix} -\frac{1}{\sqrt{2}} & \frac{1}{\sqrt{2}} \\ \frac{1}{\sqrt{2}} & \frac{1}{\sqrt{2}} \end{bmatrix}.$

Note that the resulting transformations are different.

▶ VIDEO LINK

To see simple matrix reflections and rotations explained, watch the clip at www.brightredbooks.net

 ## THINGS TO DO AND THINK ABOUT

1 (a) Write down the matrix, M_1, associated with a clockwise rotation of $\frac{\pi}{4}$ about the origin.

Express M_1 in the form $\begin{bmatrix} a & b \\ c & d \end{bmatrix}$, $a, b, c, d \in \mathbb{R}$.   2

(b) Write down a second matrix, M_2, associated with a reflection in the y-axis.   1

(c) Hence find a single matrix associated with a clockwise rotation of $\frac{\pi}{4}$ followed by a reflection in the y-axis.   2

2 Obtain the 2 × 2 matrix M associated with an enlargement, scale factor 2, followed by a clockwise rotation of 60° about the origin.   4

✅ ONLINE TEST

Test yourself on matrices at www.brightredbooks.net

COMPLEX NUMBERS

POLYNOMIAL EQUATIONS AAC

ONLINE

Find out more about complex numbers by following the links at www.brightredbooks.net

INTRODUCING COMPLEX NUMBERS

The equation $x^2 + 1 = 0$ has no solutions in $\mathbb{R}$, the set of real numbers.

The solution is $\sqrt{-1}$ and is denoted by i.

A **complex number** is of the form $z = a + bi$, where a and b are real numbers. When $b = 0$, we have the real number a.

The set of all complex numbers is denoted by $\mathbb{C}$. $\mathbb{R}$ is a subset of $\mathbb{C}$. Complex numbers obey the rules of algebra.

DON'T FORGET

$i^2 = -1,\ i^3 = -i,\ i^4 = 1$

Example: 7.1

If $z = 2 + i$, obtain z^2 and z^3.

Solution:

$z^2 = (2 + i)^2 = 4 + 4i + i^2 = 4 + 4i - 1 = 3 + 4i.$

$z^3 = z^2 z = (3 + 4i)(2 + i) = 6 + 3i + 8i + 4i^2 = 6 + 11i - 4 = 2 + 11i.$

If $z = a + bi$, the complex number $a - bi$ is called the **conjugate** of z and is denoted by $\bar{z}$.

Multiplying z by its conjugate $\bar{z}$ gives:

$z\bar{z} = (a + bi)(a - bi)$

$\quad = a^2 - abi + abi - b^2 i^2 = a^2 + b^2.$

We use this property of the complex conjugate to carry out division.

DON'T FORGET

If $z = a + bi$, then $\bar{z} = a - bi$

Example: 7.2

If $z = 2 + i$, obtain $\frac{1}{z}$ in the form $a + bi$.

Solution:

$\frac{1}{z} = \frac{1}{2 + i} = \frac{1}{2 + i} \times \frac{2 - i}{2 - i}$

$\quad = \frac{2 - i}{4 - i^2} = \frac{2 - i}{5} = \frac{2}{5} - \frac{i}{5}$

Two complex numbers $z = a + bi$ and $w = c + di$ are equal if, and only if, $a = c$ **and** $b = d$.

When working with complex numbers, you will sometimes be required to equate real and imaginary parts.

Example: 7.3

Obtain the square roots of $3 - 4i$, i.e. solve the equation $z^2 = 3 - 4i$.

Solution:

Let $z = x + yi$, so that $z^2 = x^2 + 2xyi + y^2 i^2 = x^2 - y^2 + 2xyi.$

So, $x^2 - y^2 + 2xyi = 3 - 4i \rightarrow x^2 - y^2 = 3$ and $2xy = -4$.

Hence $y = \frac{2}{x}$ and $x^2 - \frac{4}{x^2} = 3$.

Multiplying through by x^2 gives: $x^4 - 3x^2 - 4 = 0$,

and factorising gives: $(x^2 - 4)(x^2 + 1) = 0$, which gives:

$x = \pm 2$ because the second factor is always positive (remember that x is real).

$x = 2$ gives $y = -1$ and $x = -2$ gives $y = 1$.

Hence the square roots of $3 - 4i$ are $2 - i$ and $-2 + i$.

POLYNOMIAL EQUATIONS

Once we have $\mathbb{C}$, it turns out that every polynomial equation of degree n has n roots, allowing for repeated roots. This is the **fundamental theorem of algebra**.

When the coefficients are real, the complex roots (if any) occur in **conjugate pairs**.

Thus, if $3 + 2i$ is a root, then so is $3 - 2i$.

Example: 7.4

Verify that $3 + i$ is a solution of the equation $z^3 - z^2 - 20z + 50 = 0$.

Hence find all the solutions.

Solution:

$(3 + i)^2 = 8 + 6i$, $(3 + i)^3 = (8 + 6i)(3 + i) = 18 + 26i$.

Substituting into $z^3 - z^2 - 20z + 50$ gives:

$(3 + i)^3 - (3 + i)^2 - 20(3 + i) + 50 = 18 + 26i - (8 + 6i) - 60 - 20i + 50 = 0 + 0i = 0$.

Because $3 + i$ is a root, so is $3 - i$, and so:

$(z - (3 + i))(z - (3 - i)) = (z - 3 - i)(z - 3 + i) = z^2 - 6z + 10$ is a factor

of $z^3 - z^2 - 20z + 50$.

Writing $z^3 - z^2 - 20z + 50 = (z^2 - 6z + 10)(z + a)$ gives $a = 5$.

An alternative is to divide $z^3 - z^2 - 20z + 50$ by $z^2 - 6z + 10$ to give $z + 5$.

Therefore, the roots of $z^3 - z^2 - 20z + 50 = 0$ are $z = -5$, $3 + i$ and $3 - i$.

DON'T FORGET

If z is a root, then so is $\bar{z}$.

THINGS TO DO AND THINK ABOUT

1 Solve the equation $z + |z|^2 = 7 - i$, giving your answers in the form $z = a + bi$. 4

2 $z = a + 3i$

(a) Express $\frac{1}{z}$ in the form $x + iy$, where $x, y \in \mathbb{R}$. 2

(b) Given $z\bar{z} = 25$, find the possible value(s) of a. 3

3 Verify that $z = 3$ is a root of the equation
$z^3 - 9z^2 + 28z - 30 = 0$.
Hence find the other two roots. 6

4 Show that $z = 2 - 3i$ is a solution of the equation
$z^4 - 2z^3 + 7z^2 + 18z + 26 = 0$.
Hence find all the solutions. 6

5 Given that $3 - i$ is a root of the equation $2z^3 - 11z^2 + 14z + 10 = 0$,
obtain all the roots. 4

6 Given that $i - 1$ is a root of the equation $f(z) = z^4 - z^3 - 5z^2 - 8z - 2 = 0$,
obtain all the roots. 5

7 By writing $z = x + iy$, solve:

(a) $z + |z| = 8(2 - i)$

(b) $z^2 + 3 = 4(6 - 5i)$. 4

ONLINE TEST

Test yourself on polynomial equations at www.brightredbooks.net

ARGAND DIAGRAM AND GEOMETRIC FIGURES GPS

THE ARGAND DIAGRAM

The Argand diagram is a geometric way of representing complex numbers. The diagram shows a general complex number $z = a + bi$.

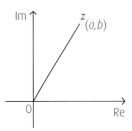

The distance of z from O in the diagram is called the **modulus** of z, denoted by $|z|$.

If $z = a + bi$, then $|z| = +\sqrt{a^2 + b^2}$.

Note that $z\bar{z} = |z|^2$.

The angle (in radians) that the line joining O to z makes with the positive x-axis is called the **argument** of z, denoted by arg z.

If $z = a + bi$, then $\arg z = \tan^{-1}\frac{b}{a}$ ($a \neq 0$).

Note that, in the third and fourth quadrants, where the imaginary part of z is negative, arg z is given as a negative angle.

If $z = bi$, then $\quad \arg z = \frac{\pi}{2}$ when $b > 0$ and $\arg z = -\frac{\pi}{2}$ when $b < 0$.

Example: 7.5

If $z = -1 + \sqrt{3}i$, show z and $\bar{z}$ on an Argand diagram. Calculate $|z|$ and arg z.

Solution:

$|z| = +\sqrt{(-1)^2 + (\sqrt{3})^2} = \sqrt{4} = 2$

From the Argand diagram, we see that arg z will be an angle between $\frac{\pi}{2}$ and π:

$\arg z = \pi - \tan^{-1}(\sqrt{3}) = \pi - \frac{\pi}{3} = \frac{2\pi}{3}$.

GEOMETRIC FIGURES IN THE COMPLEX PLANE

The expression $|z - z_0|$ gives the distance in the Argand diagram between the complex numbers z and z_0.

It follows that the equation $|z - z_0| = r$, where z_0 is fixed, and $r > 0$ is a given real number, gives all those z at a fixed distance r from z_0 – in other words, the circle, centre z_0 with radius r.

The set of all z for which $|z - z_0| < r$ gives the interior of the circle centre z_0 with radius r.

Example: 7.6

Describe the locus (the position of all points satisfying the given condition) given by the equation $|z - i| = |z + 1|$.

Solution:

Method 1

Let $z = x + yi$.
$|z - i|^2 = |x + i(y - 1)|^2 = x^2 + (y - 1)^2$.
Similarly, $|z + 1|^2 = |(x + 1) + yi|^2 = (x + 1)^2 + y^2$.
So, $x^2 + (y - 1)^2 = (x + 1)^2 + y^2$.
Expanding and simplifying gives $y = -x$, a straight line through the origin.

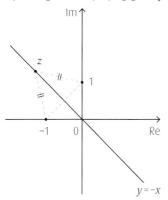

Method 2

The equation states that the distance of z from i equals its distance from -1.
Hence z lies on the perpendicular bisector of the line joining i and -1, which is the line $y = -x$.

ONLINE

For more on the topic of complex numbers, follow the link at www.brightredbooks.net

ONLINE TEST

Test yourself on complex numbers at www.brightredbooks.net

ONLINE

What are the applications of complex numbers? See online at www.brightredbooks.net

THINGS TO DO AND THINK ABOUT

1. $z = 3a + \sqrt{3}ai$, $a \in \mathbb{R}$, $a > 1$.
   - (a) Find the modulus and argument of z.   2
   - (b) Hence find z^2 and z^3 in terms of a.   3
   - (c) Plot z, z^2 and z^3 on the same Argand diagram.   3

2. The complex number z is given as $\frac{1}{\sqrt{2}} + \frac{1}{\sqrt{2}}i$.
   - (a) Show that $z^2 = i$ and $z^3 = -\frac{1}{\sqrt{2}} + \frac{1}{\sqrt{2}}i$.   2
   - (b) Plot z, z^2 and z^3 on the same Argand diagram.   3

3. Identify the locus in the complex plane given by
   - (a) $|z + 3| = 4$ 2
   - (b) $z\bar{z} = |z - 2i|^2$ 4

4. Show, on an Argand diagram, the locus of all points satisfying the inequality $|z - 2i| < 3$.   2

5. Identify the locus in the complex plane given by $|z - 2i| = |z + 6|$.   3

POLAR FORM AND DE MOIVRE'S THEOREM GPS

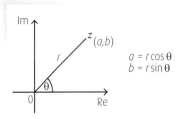

We can also use polar coordinates to represent complex numbers.

In this form, $z = r(\cos\theta + i\sin\theta)$ where $r = |z|$ and $\theta = \arg z$.

Products are easy in this form:

If $z_1 = r_1(\cos\theta_1 + i\sin\theta_1)$ and $z_2 = r_2(\cos\theta_2 + i\sin\theta_2)$, then:

$z_1 z_2 = r_1 r_2 (\cos(\theta_1 + \theta_2) + i\sin(\theta_1 + \theta_2))$.

This shows that $|z_1 z_2| = |z_1||z_2|$ and $\arg(z_1 z_2) = \arg(z_1) + \arg(z_2)$ (with a possible adjustment if this sum lies outside the range $-\pi < \theta \leqslant \pi$).

We note also that, when the argument is negative, the complex number z can be written in a slightly different way.

Using trig properties of negative angles:

$z = r(\cos(-\theta) + i\sin(-\theta)) = r(\cos\theta - i\sin\theta)$.

If we take $z = \cos\theta + i\sin\theta$, then $z^n = \cos n\theta + i\sin n\theta$.

This can be written as: $(\cos\theta + i\sin\theta)^n = \cos n\theta + i\sin n\theta$.

This is **de Moivre's theorem**. It is valid for any integer n.

DON'T FORGET

If $z = r(\cos\theta + i\sin\theta)$, then $z^n = r^n(\cos n\theta + i\sin n\theta)$.

ONLINE

To find out more about de Moivre himself, go to www.brightredbooks.net

Example: 7.7

Express $\sqrt{3} + i$ in polar form and hence express $(\sqrt{3} + i)^8$ in the form $a + bi$.

Solution:

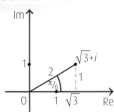

We have $\sqrt{3} + i = 2\left(\cos\frac{\pi}{6} + i\sin\frac{\pi}{6}\right)$, so:

$(\sqrt{3} + i)^8 = 2^8\left(\cos\frac{8\pi}{6} + i\sin\frac{8\pi}{6}\right)$.

$\frac{8\pi}{6} = \frac{4\pi}{3}$ and $\cos\frac{4\pi}{3} = -\frac{1}{2}$, $\sin\frac{4\pi}{3} = -\frac{\sqrt{3}}{2}$, so:

$(\sqrt{3} + i)^8 = 2^8\left(-\frac{1}{2} - \frac{\sqrt{3}}{2}i\right) = -2^7 - 2^7 i\sqrt{3}$.

Example: 7.8

Use de Moivre's theorem to prove that:

$\cos 3\theta = 4\cos^3\theta - 3\cos\theta$ and $\sin 3\theta = 3\sin\theta - 4\sin^3\theta$.

contd

Solution:

Using de Moivre's theorem, we know $(\cos \theta + i \sin \theta)^3 = \cos 3\theta + i \sin 3\theta$.

Also, expanding the left-hand side by the binomial theorem:

$(\cos \theta + i \sin \theta)^3 = \cos^3 \theta + 3i \cos^2 \theta \sin \theta + 3i^2 \cos \theta \sin^2 \theta + i^3 \sin^3 \theta$.

Because $i^2 = -1$, $i^3 = -i$, this simplifies to:

$(\cos \theta + i \sin \theta)^3 = \cos^3 \theta + 3i \cos^2 \theta \sin \theta - 3 \cos \theta \sin^2 \theta - i \sin^3 \theta$.

Hence:

$\cos 3\theta + i \sin 3\theta = \cos^3 \theta - 3 \cos \theta \sin^2 \theta + 3i \cos^2 \theta \sin \theta - i \sin^3 \theta$.

Equating the real parts on both sides, and also the imaginary parts:

$\cos 3\theta = \cos^3 \theta - 3 \cos \theta \sin^2 \theta$ and $\sin 3\theta = 3 \cos^2 \theta \sin \theta - \sin^3 \theta$.

Finally, using $\cos^2 \theta + \sin^2 \theta = 1$ gives the results:

$\cos 3\theta = 4 \cos^3 \theta - 3 \cos \theta$ and $\sin 3\theta = 3 \sin \theta - 4 \sin^3 \theta$.

Of course, these results can be obtained without using complex numbers, but many such results are best proved in this way.

nTH ROOTS OF COMPLEX NUMBERS

If z is an nth root of a given complex number z_0, then $z^n = z_0$. By the fundamental theorem of algebra, this equation has n roots, so every complex number has n nth roots (remember, this includes real numbers as well).

Example: 7.9

Use de Moivre's theorem to obtain the cube roots of –1.

Solution:

We must solve the equation $z^3 = -1$, so begin by setting $z = r (\cos \theta + i \sin \theta)$.

Then $z^3 = r^3 (\cos 3\theta + i \sin 3\theta)$, so $r^3 \cos 3\theta + i r^3 \sin 3\theta = -1$.

The modulus of the left-hand side is r^3, and for the right-hand side it is 1. Hence $r = 1$.

So, $\cos 3\theta + i \sin 3\theta = -1$, and equating real and imaginary parts on both sides gives $\cos 3\theta = -1$ and $\sin 3\theta = 0$.

The values for which $\cos 3\theta = -1$ **and** $\sin 3\theta = 0$ **where also** $-\pi < \theta \leq \pi$ are:

$\theta = \pi$, $\theta = \frac{\pi}{3}$, $\theta = -\frac{\pi}{3}$.

So, $z = \cos \pi + i \sin \pi$, $z = \cos\frac{\pi}{3} + i \sin\frac{\pi}{3}$, $z = \cos\frac{\pi}{3} - i \sin\frac{\pi}{3}$,

giving $z = -1$, $z = \frac{1}{2} + \frac{\sqrt{3}}{2}i$, $z = \frac{1}{2} - \frac{\sqrt{3}}{2}i$.   Note that the complex roots are conjugates.

ONLINE

Follow the link at www.brightredbooks.net. This webpage states and proves **de Moivre's theorem** and presents several examples of its use.

VIDEO LINK

Check out the clip at www.brightredbooks.net

THINGS TO DO AND THINK ABOUT

ONLINE TEST

Test yourself on this topic at www.brightredbooks.net

1 Given $z = 2 + ai$, where $a > 0$, obtain the value of a given that z^3 is real. Evaluate $|z|$ and arg z.   6

2 (a) Use de Moivre's theorem to express z^k in terms of θ, where $z = \cos \theta + i \sin \theta$, and k is a positive integer.   1

   (b) Hence show that $z^{-k} = \cos k\theta - i \sin k\theta$, and obtain expressions for $\cos k\theta$ and $\sin k\theta$ in terms of z.   2

   (c) Show that $\cos 2\theta \sin^2 \theta = -\frac{1}{8}\left(z^2 + \frac{1}{z^2}\right)\left(z - \frac{1}{z}\right)^2$.   2

   (d) Hence show that $\cos 2\theta \sin^2 \theta = a + b \cos 2\theta + c \cos 4\theta$ for suitable constants a, b, c.   3

3 (a) Use the binomial theorem to obtain the full expansion of $(\cos x + i \sin x)^3$.   2

   (b) Use de Moivre's theorem to obtain a second expression for $(\cos x + i \sin x)^3$.   1

   (c) Hence obtain an expression for $\cos 3x$ in terms of $\cos x$ only.   3

SEQUENCES AND SERIES

ARITHMETIC SEQUENCES AAC

OVERVIEW

DON'T FORGET

The nth term of the arithmetic sequence with first term a and common difference d is:
$u_n = a + (n - 1)d$

In an arithmetic sequence, consecutive terms change by a constant amount:
$u_1 = a$, $u_2 = a + d$, $u_3 = a + 2d$, ..., and generally: $u_n = a + (n - 1)d$.

Note that $u_{k+1} = u_k + d$, where d is the **common difference** (it may be positive or negative). The associated series $S_n = a + (a + d) + (a + 2d) + ... + (a + [n - 1]d)$ is called an **arithmetic series**.

Example: 8.1

The sequence 3, 5, 7, ... is an arithmetic sequence with $a = 3$ and $d = 2$.
We have: $u_n = 2n + 1$ ($n = 1, 2, 3, ...$).

DON'T FORGET

In an arithmetic sequence, add d to get the next term.

SUM OF AN ARITHMETIC SERIES

Let $S_n = a + (a + d) + (a + 2d) + ... + (a + [n - 1]d)$

$= an + d(1 + 2 + 3 + ... + (n - 1))$

$= an + \frac{1}{2}n(n - 1)d$

$= \frac{n}{2}\{2a + (n - 1)d\}$

This formula should be remembered, although it does appear in the formulae sheet. An alternative form, which is sometimes useful, is:

$S_n = \frac{n}{2}(u_1 + u_n)$, where u_1 and u_n are the first and last terms in the sequence.

DON'T FORGET

$1 + 2 + 3 + ... + n = \frac{n(n + 1)}{2}$

Example: 8.2

Evaluate $\sum_{k=1}^{6} 4$.

Solution:

$\sum_{k=1}^{6} 4 = 6 \times 4 = 24$

DON'T FORGET

$\sum_{k=1}^{n} c = nc$ for any constant c.

Example: 8.3

Obtain a formula for $\sum_{k=1}^{n} (3k + 1)$.   Hence evaluate $301 + 304 + 307 + ... + 601$.

Solution:

Method 1

This is an arithmetic series with $a = 4$ and $d = 3$.
Writing out the first few terms, i.e. 4, 7, 10, ..., will make this clear.
We have $a = 4$, $d = 3$, so:
$S_n = \frac{n}{2}\{8 + 3(n - 1)\} = \frac{1}{2}n(3n + 5)$.
Hence: $301 + 304 + 307 + ... + 601 = S_{200} - S_{99}$ since $k = 200$ gives the 601 term and $k = 99$ gives the term before 301.

So: $301 + 304 + 307 + ... + 601 = \frac{200 \times 605}{2} - \frac{99 \times 302}{2} = 45\,551$.
Alternatively, using $S_n = \frac{n}{2}(u_1 + u_n)$, we have $200 - 99 = 101$ terms and:

$301 + 304 + 307 + ... + 601 = \frac{101}{2}(301 + 601) = 45\,551$.

Method 2
$\sum_{k=1}^{n} (3k + 1) = 3\sum_{k=1}^{n} k + \sum_{k=1}^{n} 1 = \frac{3n(n + 1)}{2} + n$

$= \frac{n}{2}(3n + 3 + 2) = \frac{n}{2}(3n + 5)$ Evaluation as method 1.

contd

Example: 8.4

(i) Obtain a formula for the sum, S_n, of the first n terms of the sequence 9, 7, 5, ...

(ii) Find the values of n for which $S_n = 21$.

(iii) Obtain the smallest value of n for which $S_n < -50$.

Solution:

(i) The sequence is arithmetic with $a = 9$ and $d = -2$, so:
$$S_n = \frac{n}{2}\{18 - 2(n - 1)\} = \frac{n}{2}(20 - 2n) = 10n - n^2$$

(ii) $S_n = 21 \Rightarrow 10n - n^2 = 21$, so $n^2 - 10n + 21 = 0$

$(n - 3)(n - 7) = 0$, so the sum is 21 when $n = 3$ and when $n = 7$.

(iii) We require the smallest value of n ($n \geq 1$) for which $10n - n^2 < -50 \Rightarrow n^2 - 10n > 50$.

Completing the square, $(n - 5)^2 > 75$, so $n > 5 + \sqrt{75} \approx 13 \cdot 7$.

So, the smallest value of n is 14.

(An alternative to completing the square is to solve $x^2 - 10x - 50 = 0$, using the formula, and explain why (with a rough diagram, for example) the value of n is the smallest integer greater than the positive root of this quadratic.)

 **DON'T FORGET**

Sum (arithmetic)
$\frac{n}{2}[2a + (n - 1)d]$

 THINGS TO DO AND THINK ABOUT

1 The first term of an arithmetic sequence is 3, and the tenth term is –15. Obtain the sum of the first 100 terms. **4**

2 Show that $\sum_{r=1}^{n} (6 + 2r) = 7n + n^2$. **2**

Hence write down a formula for $\sum_{r=1}^{3k} (6 + 2r)$. **1**

Show that $\sum_{r=k+1}^{3k} (6 + 2r) = 14k + 8k^2$. **2**

3 The sum to n terms of a sequence is given by $S_n = 2n - n^2$, $n \in \mathbb{N}$.

(a) Obtain the first three terms of the sequence: u_1, u_2 and u_3. **2**

(b) Obtain a formula for u_n. **3**

 ONLINE

Head to www.brightredbooks.net to explore this topic further.

 ONLINE TEST

Test yourself on arithmetic sequences at www.brightredbooks.net

GEOMETRIC SEQUENCES AAC

In a geometric sequence, consecutive terms change by a constant factor: $a, ar, ar^2, \ldots$

The factor r is called the **common ratio**, and can have any value (but when $r = 0$ or $r = 1$ the sequence becomes trivial).

Note that $u_k = ar^{k-1}$ and $u_{k+1} = ru_k$.

The partial sums of geometric sequences are called **geometric series**.

Example: 8.5

Write down the general term of the geometric sequence $5, -10, 20, \ldots$

$a = 5$ and $r = -2$, so $u_k = 5(-2)^{k-1} = (-1)^{k-1}5 \times 2^{k-1}$.

In any geometric sequence, $r = \dfrac{u_2}{u_1} = \dfrac{u_3}{u_2} = \ldots = \dfrac{u_n}{u_{n-1}}$.

This can be used to test whether a number sequence is in fact geometric, and also to find the value of r if you cannot spot it easily.

SUM OF A GEOMETRIC SERIES

The nth partial sum is $a + ar + ar^2 + \ldots + ar^{n-1} = \dfrac{a(1 - r^n)}{1 - r} = \dfrac{a(r^n - 1)}{r - 1}$ $(r \neq 1)$.

Example: 8.6

Obtain the sum of the first 20 terms of the geometric sequence $3, 1, \frac{1}{3}, \ldots$

Solution:

We have $a = 3$ and $r = \frac{1}{3}$, so the sum of 20 terms is

$\dfrac{3\left(1 - \left(\frac{1}{3}\right)^{20}\right)}{1 - \frac{1}{3}} = \dfrac{9}{2}(1 - 3^{-20})$.

SUM TO INFINITY

When $-1 < r < 1$, $r^n \to 0$ as $n \to \infty$, and so:

$\dfrac{a(1 - r^n)}{1 - r} \to \dfrac{a}{1 - r}$.

This is called the **sum to infinity**, and we write:

$a + ar + ar^2 + \ldots = \displaystyle\sum_{n=1}^{\infty} ar^{n-1} = \dfrac{a}{1 - r}$.

Example: 8.7

$1 + \frac{1}{2} + \frac{1}{4} + \frac{1}{8} + \ldots = \dfrac{1}{1 - \frac{1}{2}} = 2$

Example: 8.8

For what values of x does the infinite series $1 + 2x^2 + 4x^4 + 8x^6 + \ldots$ have a sum, and what is this sum?

Solution:

This is a geometric series with $a = 1$ and $r = 2x^2$, so we need $-1 < 2x^2 < 1$.

The lower limit is automatically true, and the upper limit gives:

$-\dfrac{1}{\sqrt{2}} < x < \dfrac{1}{\sqrt{2}}$.

The sum is given by:

$\dfrac{a}{1 - r} = \dfrac{1}{1 - 2x^2}$.

MORE SUMS

Example: 8.9

Evaluate the sum $\sum_{r=10}^{20} (3r - 5)$.

Solution:

Method 1

$$\sum_{r=10}^{20} (3r - 5) = \left(3\sum_{r=1}^{20} r - \sum_{r=1}^{20} 5\right) - \left(3\sum_{r=1}^{9} r - \sum_{r=1}^{9} 5\right)$$

$$= \left(\tfrac{3}{2} \times 20(21) - 5 \times 20\right) - \left(\tfrac{3}{2} \times 9(10) - 5 \times 9\right)$$

$$= (630 - 100) - (135 - 45)$$

$$= (530 - 90)$$

$$= 440$$

Method 2

$$\sum_{r=10}^{20} (3r - 5) = \sum_{r=1}^{20}(3r - 5) - \sum_{r=1}^{9}(3r - 5)$$

$$= (-2 + 1 + 4 + \ldots + 55) - (-2 + 1 + 4 + \ldots + 22)$$

$$= \tfrac{20}{2}(-2 + 55) - \tfrac{9}{2}(-2 + 22) = 530 - 90 = 440$$

The square and cube numbers do not form arithmetic or geometric sequences. However, the formula for the sum of their terms is known. Remember these results:

$$\sum_{k=1}^{n} k = \frac{n(n + 1)}{2} \qquad \sum_{k=1}^{n} k^2 = \frac{n(n + 1)(2n + 1)}{6} \qquad \sum_{k=1}^{n} k^3 = \left(\frac{n(n + 1)}{2}\right)^2$$

 ## THINGS TO DO AND THINK ABOUT

1   The third term of a geometric series is $1\frac{1}{4}$ and the eighth term is $\frac{5}{128}$. Determine the sum of the first eight terms of the series.   **4**

2   The second and fourth terms of a geometric series are -2 and $-\frac{2}{9}$ respectively. Explain why the series has a sum to infinity, and obtain two possible values for this sum.   **5**

3   An arithmetic progression and a geometric progression have the same first term, 4, and the same sum to three terms.
One possible value for the common ratio, r, of the geometric progression is 2. Find the other possible value for r, and the common difference of the arithmetic progression.   **5**

4   An arithmetic sequence has a sum to seven terms of $66\frac{1}{2}$. The seventh term is 16. Find u_3.   **5**

MACLAURIN SERIES $\overline{\text{AAC}}$

When you use your calculator to find the value of sin 40°, how can it give an answer to such accuracy? Certainly not from any scale drawing involving a 40° triangle.

Here is one of the methods which are used to evaluate accurately advanced formulae such as trigonometric, exponential and logarithmic functions.

Suppose a function $f(x)$ has derivatives of all orders defined at $x = 0$. Denote the rth derivative evaluated at $x = 0$ by $f^{(r)}(0)$ for $r = 1, 2, 3, ...$, and form the infinite series:

$$f(x) = f(0) + f'(0)x + \frac{f''(0)x^2}{2!} + \frac{f'''(0)x^3}{3!} + \frac{f^{iv}(0)x^4}{4!} + ...$$

DON'T FORGET

The Maclaurin series of $f(x)$ is $f(0) + \sum\limits_{r=1}^{\infty} \frac{f^{(r)}(0)x^r}{r!}$

When this series converges (the values of x may be restricted for this), it is called the Maclaurin series (or expansion) of $f(x)$, after the Scottish mathematician Colin Maclaurin (1698–1746), who did much early work on these series.

The sum is $f(x)$ when the series converges for all the functions you will meet at Advanced Higher.

Example: 8.10

Find the Maclaurin series for $f(x) = \frac{1}{1-x}$.

Solution:

We have $f(0) = 1$, $f'(0) = 1$, $f''(0) = 2$, and in general $f^{(r)}(0) = r!$ (verify a few more derivatives to see the pattern).

So, the Maclaurin series for

$\frac{1}{1-x}$ is $1 + x + x^2 + ...$

From the section on geometric series, we know that this series only converges for $-1 < x < 1$, so, although the function $f(x) = \frac{1}{1-x}$ is defined for all $x \neq 1$, its Maclaurin series is only defined for $-1 < x < 1$.

Example: 8.11

Obtain the Maclaurin series for $f(x) = (1 + x)^n$ where n is a positive integer.

Solution:

$f'(x) = n(1 + x)^{n-1}$, $f''(x) = n(n - 1)(1 + x)^{n-2}$, and generally:

$f^{(r)}(x) = n(n - 1) ... (n - r + 1)(1 + x)^{n-r}$ for $r \leq n$. For $r > n$, all derivatives are zero.

So, the Maclaurin series for $(1 + x)^n$ is:

$1 + nx + \frac{n(n-1)}{2!}x^2 + ... + \frac{n(n-1)...(n-r+1)}{r!}x^r + ... + x^n$

$= 1 + \binom{n}{1}x + \binom{n}{2}x^2 + ... + \binom{n}{r}x^r + ... + \binom{n}{n}x^n$.

Because this is a finite series (polynomial), it is valid for all x. We recognise this as the binomial expansion from Chapter 1: Algebra. Now we can use Maclaurin series to expand binomials with fractional and negative powers.

Example: 8.12

Obtain the first four terms in the Maclaurin series for $\sqrt{1 + x}$.

contd

Solution:

$f(x) = \sqrt{1 + x}$, $f(0) = 1$

$f'(x) = \frac{1}{2}(1 + x)^{-\frac{1}{2}}$, $f'(0) = \frac{1}{2}$

$f''(x) = \left(\frac{1}{2}\right)\left(-\frac{1}{2}\right)(1 + x)^{-\frac{3}{2}}$, $f''(0) = -\frac{1}{4}$

$f'''(x) = \left(\frac{1}{2}\right)\left(-\frac{1}{2}\right)\left(-\frac{3}{2}\right)(1 + x)^{-\frac{5}{2}}$, $f'''(0) = \frac{3}{8}$.

So, for the first four terms, we have:

$\sqrt{1 + x} = 1 + \frac{x}{2} - \frac{x^2}{8} + \frac{x^3}{16} + \dots$

The series converges for $-1 < x < 1$ (you are not expected to prove this), although the function is defined for all $x > -1$.

The following table gives some standard results which are worth remembering.
You are expected to know how to derive them.

Function	Maclaurin series	valid for
e^x	$1 + x + \frac{x^2}{2!} + \frac{x^3}{3!} + \dots$	all x
$\sin x$	$x - \frac{x^3}{3!} + \frac{x^5}{5!} - \dots$	all x
$\cos x$	$1 - \frac{x^2}{2!} + \frac{x^4}{4!} - \dots$	all x
$\ln(1 + x)$	$x - \frac{x^2}{2} + \frac{x^3}{3} - \dots$	$-1 < x \leqslant 1$

Let's think about this

Why does the Maclaurin series for $\cos x$ only contain even powers, while that for $\sin x$ only contains odd powers?

Answer: $\cos x$ is an even function of x, and $\sin x$ is an odd function of x.

Example: 8.13

Obtain the first three terms in the Maclaurin series for $\cos 5x$.

Solution:

The standard result for $\cos x$ may be used and then $5x$ substituted for x.

$\cos x = 1 - \frac{x^2}{2!} + \frac{x^4}{4!}$

Hence $\cos 5x = 1 - \frac{(5x)^2}{2!} + \frac{(5x)^4}{4!} = 1 - \frac{25x^2}{2} + \frac{625x^4}{24}$

 THINGS TO DO AND THINK ABOUT

1 Use Maclaurin's theorem to obtain the series expansion of $\tan x$ up to and including the term in x^4. **4**

2 Use Maclaurin's theorem to obtain the first three non-zero terms of the series expansion for $\ln(x + 2)$, $-2 < x \leqslant 2$. **3**

3 Deduce the Maclaurin series for $\sin 3x$ as far as the term in x^5. **2**
Hence obtain the Maclaurin series for $(1 + x)\sin 3x$ as far as the term in x^5. **1**

 DON'T FORGET

Try to use standard results as much as possible.

 **ONLINE**

Follow the links at www.brightredbooks.net for more on Maclaurin and Taylor series.

 VIDEO LINK

Watch the clip at www.brightredbooks.net to see another example.

 ONLINE TEST

Test yourself on geometric sequences at www.brightredbooks.net

ASSOCIATED SERIES

Using the table on page 65, we can obtain Maclaurin expansions for more complicated functions.

Example: 8.14

Obtain the Maclaurin series for $\sin^2 x$ as far as the term in x^4.

Solution:

Note that, if a question hasn't directed you to use a particular method, any of these methods would be acceptable.

Method 1

$\sin^2 x = \frac{1}{2}(1 - \cos 2x)$

$\qquad = \frac{1}{2} - \frac{1}{2}\left(1 - \frac{(2x)^2}{2!} + \frac{(2x)^4}{4!} - \ldots\right)$

$\qquad = \frac{1}{2} - \frac{1}{2} + \frac{4x^2}{4} - \frac{16x^4}{48} + \ldots$

$\qquad = x^2 - \frac{x^4}{3} + \ldots$

Method 2

$f(x) = \sin^2 x,\ f(0) = 0$

$f'(x) = 2\sin x \cos x = \sin 2x,\ f'(0) = 0.$

Using the double angle formula here means that we do not require the product rule for further derivatives.

$f''(x) = 2\cos 2x,\ f''(0) = 2$

$f'''(x) = -4\sin 2x,\ f'''(0) = 0$

$f^{iv}(x) = -8\cos 2x,\ f^{iv}(0) = -8.$

Check that these results give the same answer as before.

Method 3

$\sin^2 x = \sin x \sin x = \left(x - \frac{x^3}{6} + \ldots\right)\left(x - \frac{x^2}{6} + \ldots\right)$ (ignoring higher powers)

$\qquad = x^2 - \frac{x^4}{6} - \frac{x^4}{6} + \ldots,$ giving the same result as before.

Example: 8.15

Obtain the Maclaurin series for $f(x) = xe^{-x^2}$ as far as the term in x^5.

Solution:

Replacing x by $-x^2$ in the expansion for e^x gives:

$e^{-x^2} = 1 - x^2 + \frac{x^4}{2} - \ldots,\quad$ so: $f(x) = x - x^3 + \frac{x^5}{2} - \ldots$

To differentiate the function xe^{-x^2} five times using the product and chain rules would involve far more work than the marks awarded (probably about 3 or 4) would justify.

THINGS TO DO AND THINK ABOUT

1　Obtain the first three non-zero terms in the Maclaurin series of $\ln(3 - x)$. 　3

　　Hence, or otherwise, obtain the first three non-zero terms in the Maclaurin series of $x^2 \ln(3 - x)$ and $x^2 \ln(3 + x)$. 　2

　　Hence obtain the first **two** non-zero terms in the Maclaurin expansion of $x^2 \ln(9 - x^2)$. 　2

2　Obtain the Maclaurin expansion of $e^{2x}\cos 3x$ up to and including the term in x^3. 　6

DON'T FORGET

Be prepared to combine Maclaurin series with other algebra or trigonometry topics.

BINOMIAL EXPANSION WHERE THE INDEX, n, IS A NEGATIVE INTEGER

Consider $1 - x + x^2 - x^3 \ldots$

This is a **geometric series** with first term 1 and common ratio $-x$.
Provided $|x| < 1$ (alternatively write $-1 < x < 1$), a sum to infinity exists.

$$S_\infty = \frac{1}{1 - (-x)} = \frac{1}{1 + x} = (1 + x)^{-1}$$

so, $(1 + x)^{-1} = 1 - x + x^2 - x^3 \ldots$ for $-1 < x < 1$

which is the binomial expansion for $n = -1$.

We need to note the series expansion does not terminate but is an infinite series and its sum will never be exactly $(1 + x)^{-1}$.

Remember $\quad (1 + x)^n = {}^nC_0 1^n x^0 + {}^nC_1 1^{n-1} x^1 + {}^nC_2 1^{n-2} x^2 + \ldots$

$$= 1 \times 1 \times x^0 + \frac{n!}{(n-1)!1!} \times 1 \times x^1 + \frac{n!}{(n-2)!2!} \times 1 \times x^2 + \frac{n!}{(n-3)!3!} \times 1 \times x^3 + \ldots$$

$$(1 + x)^n = 1 + nx + \frac{n(n-1)}{2}x^2 + \frac{n(n-1)(n-2)}{3!}x^3 + \ldots$$

DON'T FORGET

For negative indices, expression must be of the form $p(1 \pm qx)^{-k}$, q > 0.

DON'T FORGET

Expansion valid for $-1 < qx < 1$
so $-\frac{1}{q} < x < \frac{1}{q}$,
$q > 0$.

Example: 8.16

Expand $\frac{1}{(1 - 3x)^2}$ up to and including the term in x^3, and state the values of x for which it is valid.

Solution:

$\frac{1}{(1 - 3x)^2} = 1 + (-2)(-3x) + \frac{(-2)(-3)}{2}(-3x)^2 + \frac{(-2)(-3)(-4)}{3!}(-3x)^3 \ldots$

provided $-1 < 3x < 1$

$-\frac{1}{3} < x < \frac{1}{3}$

VIDEO LINK

See more on this topic, including negative and fractional powers, at www.brightredbooks.net

For $(a + x)^n$, write as $\left(a\left(1 + \frac{x}{a}\right)\right)^n = a^n\left(1 + \frac{x}{a}\right)^n$ since the expansion is only valid for $(1 + x)^n$.

Example: 8.17

Expand $\frac{1}{(4 - x)^2}$ up to and including the term in x^3.

Solution:

$\frac{1}{(4 - x)^2} = (4 - x)^{-2} = \left(4\left(1 - \frac{x}{4}\right)\right)^{-2} = 4^{-2}\left(1 - \frac{x}{4}\right)^{-2}$

$4^{-2}\left(1 - \frac{x}{4}\right)^{-2} = 4^{-2}\left[1 + (-2)\left(-\frac{x}{4}\right) + \frac{(-2)(-3)}{2}\left(-\frac{x}{4}\right)^2 + \frac{(-2)(-3)(-4)}{3!}\left(-\frac{x}{4}\right)^3 + \ldots\right]$

$= \frac{1}{16}\left[1 + \frac{x}{2} + \frac{3x^2}{16} + \frac{x^3}{16}\right] = \frac{1}{16} + \frac{x}{32} + \frac{3x^2}{256} + \frac{x^3}{256}.$

Valid for $-1 < \frac{x}{4} < 1$

So $\quad -4 < x < 4$

THINGS TO DO AND THINK ABOUT

1 Expand $\frac{6}{(1 + 2x)(1 - 4x)}$ up to and including the term in x^3.

This could be done by writing the expression in partial fractions and then producing separate series – you should try this method and also method 2, writing as $6(1 + y)^{-1}$ where $y = -2x - 8x^2$.   **8**

2 Express $f(x) = \frac{x^2 + 8x + 11}{(x + 1)(x + 3)^2}$ $(x \neq -1, -3)$ in partial fractions.   **4**

Hence obtain the first three non-zero terms in the Maclaurin expansion of $f(x)$.
For what values of x does this series converge?   **5**

ONLINE TEST

Test yourself on geometric sequences at www.brightredbooks.net

VECTORS

BASIC SKILLS INFORMATION GPS

Vectors have magnitude and direction. Examples are velocity, force and displacements in three-dimensional space. Numbers are called **scalars** to distinguish them from vectors.

We denote vectors by boldface type, e.g. **u**, but in your written work you should underline vectors, e.g. u̲.

Unit vectors have magnitude 1. The unit displacement vectors along the positive Ox-, Oy- and Oz-axes in three-dimensional space are denoted by **i**, **j** and **k**.

For a point P with coordinates (a, b, c), the position vector $\overrightarrow{OP}$ is the displacement from O to P $a\boldsymbol{i} + b\boldsymbol{j} + c\boldsymbol{k}$, where '+' means a units along the x-axis, followed by b units along the y-axis, then c units along the z-axis.

Another way to represent the vector $a\boldsymbol{i} + b\boldsymbol{j} + c\boldsymbol{k}$ is the column matrix $\begin{bmatrix} a \\ b \\ c \end{bmatrix}$.

This is called a **column vector**.

The scalars a, b and c are called the **components** of the vector $a\boldsymbol{i} + b\boldsymbol{j} + c\boldsymbol{k}$.

The vector $\begin{bmatrix} 0 \\ 0 \\ 0 \end{bmatrix}$ is the **zero vector**, denoted by **0**.

If $\boldsymbol{p} = a\boldsymbol{i} + b\boldsymbol{j} + c\boldsymbol{k}$, then for any scalar k:

$k\boldsymbol{p} = ka\boldsymbol{i} + kb\boldsymbol{j} + kc\boldsymbol{k}$.

The magnitude of the vector $a\boldsymbol{i} + b\boldsymbol{j} + c\boldsymbol{k}$ is $\sqrt{a^2 + b^2 + c^2}$ (length and magnitude are always $\geqslant 0$), so the unit vectors pointing in the direction determined by $\overrightarrow{OP}$ are given by:

$+\dfrac{a\boldsymbol{i} + b\boldsymbol{j} + c\boldsymbol{k}}{\sqrt{a^2 + b^2 + c^2}}$ and $-\dfrac{a\boldsymbol{i} + b\boldsymbol{j} + c\boldsymbol{k}}{\sqrt{a^2 + b^2 + c^2}}$.

The minus sign gives the unit vector in the direction $\overrightarrow{PO}$.

Addition is defined by:

$(a_1\boldsymbol{i} + b_1\boldsymbol{j} + c_1\boldsymbol{k}) + (a_2\boldsymbol{i} + b_2\boldsymbol{j} + c_2\boldsymbol{k}) = (a_1 + a_2)\boldsymbol{i} + (b_1 + b_2)\boldsymbol{j} + (c_1 + c_2)\boldsymbol{k}$.

Example: 9.1
If $\boldsymbol{a} = 2\boldsymbol{i} + \boldsymbol{j} - 3\boldsymbol{k}$ and $\boldsymbol{b} = \boldsymbol{i} - 2\boldsymbol{j} + \boldsymbol{k}$, obtain $3\boldsymbol{a} - 2\boldsymbol{b}$.

Solution:
$3\boldsymbol{a} - 2\boldsymbol{b} = 3(2\boldsymbol{i} + \boldsymbol{j} - 3\boldsymbol{k}) - 2(\boldsymbol{i} - 2\boldsymbol{j} + \boldsymbol{k}) = 4\boldsymbol{i} + 7\boldsymbol{j} - 11\boldsymbol{k}$.

VECTOR PRODUCTS

Scalar product (or dot product)

Given two vectors $\boldsymbol{a}$ and $\boldsymbol{b}$, we define the **scalar product** (or dot product) of $\boldsymbol{a}$ and $\boldsymbol{b}$ by:

$\boldsymbol{a}.\boldsymbol{b} = |\boldsymbol{a}|\,|\boldsymbol{b}|\cos\theta$.

Note that:

$\boldsymbol{a}.\boldsymbol{b} = \boldsymbol{b}.\boldsymbol{a}$ and $\boldsymbol{a}.(\boldsymbol{b} + \boldsymbol{c}) = \boldsymbol{a}.\boldsymbol{b} + \boldsymbol{a}.\boldsymbol{c}$.

When $\theta = 90°$, $\cos\theta = 0$, so:

$\boldsymbol{a}.\boldsymbol{b} = 0$ if and only if $\boldsymbol{a}$ and $\boldsymbol{b}$ are perpendicular.

We have $\boldsymbol{i}.\boldsymbol{j} = \boldsymbol{j}.\boldsymbol{k} = \boldsymbol{k}.\boldsymbol{i} = 0$, and $\boldsymbol{i}.\boldsymbol{i} = \boldsymbol{j}.\boldsymbol{j} = \boldsymbol{k}.\boldsymbol{k} = 1$. Using these results, we have:

$(a_1\boldsymbol{i} + b_1\boldsymbol{j} + c_1\boldsymbol{k}).(a_2\boldsymbol{i} + b_2\boldsymbol{j} + c_2\boldsymbol{k}) = a_1a_2 + b_1b_2 + c_1c_2$.

contd

Example: 9.2

For what value of a are the vectors $2\boldsymbol{i} + \boldsymbol{j} + \boldsymbol{k}$ and $\boldsymbol{i} + a\boldsymbol{j} - 3\boldsymbol{k}$ perpendicular to each other?

Solution:

$(2\boldsymbol{i} + \boldsymbol{j} + \boldsymbol{k}).(\boldsymbol{i} + a\boldsymbol{j} - 3\boldsymbol{k}) = 2 + a - 3 = 0$, so $a = 1$.

Cross product

As well as the scalar product, we can also define a product of two vectors which gives a vector: $\boldsymbol{a} \times \boldsymbol{b}$ (read as '$\boldsymbol{a}$ cross $\boldsymbol{b}$') is a vector with magnitude $|\boldsymbol{a}||\boldsymbol{b}| \sin \theta$ pointing as shown in the diagram.

Note that $\boldsymbol{a} \times \boldsymbol{b} = -\boldsymbol{b} \times \boldsymbol{a}$ and $\boldsymbol{a} \times \boldsymbol{a} = \boldsymbol{0}$.

Also $\boldsymbol{a} \times (\boldsymbol{b} + \boldsymbol{c}) = \boldsymbol{a} \times \boldsymbol{b} + \boldsymbol{a} \times \boldsymbol{c}$.

We have $\boldsymbol{i} \times \boldsymbol{j} = \boldsymbol{k}$, $\boldsymbol{j} \times \boldsymbol{k} = \boldsymbol{i}$ and $\boldsymbol{k} \times \boldsymbol{i} = \boldsymbol{j}$.

Applying the above rules, we get:

$(a_1\boldsymbol{i} + b_1\boldsymbol{j} + c_1\boldsymbol{k}) \times (a_2\boldsymbol{i} + b_2\boldsymbol{j} + c_2\boldsymbol{k}) = (b_1c_2 - b_2c_1)\boldsymbol{i} - (a_1c_2 - a_2c_1)\boldsymbol{j} + (a_1b_2 - a_2b_1)\boldsymbol{k}$.

This is best remembered as a determinant:

$$\begin{vmatrix} \boldsymbol{i} & \boldsymbol{j} & \boldsymbol{k} \\ a_1 & b_1 & c_1 \\ a_2 & b_2 & c_2 \end{vmatrix}$$

The two types of product can be combined in various ways.

 DON'T FORGET

Scalar product, . , = Scalar.
Vector product, × , = Vector.

Example: 9.3

If $\boldsymbol{a} = \boldsymbol{i} + 2\boldsymbol{j} - \boldsymbol{k}$, $\boldsymbol{b} = 2\boldsymbol{i} - \boldsymbol{j} + 2\boldsymbol{k}$ and $\boldsymbol{c} = \boldsymbol{i} + \boldsymbol{j}$, obtain $\boldsymbol{a}.(\boldsymbol{b} \times \boldsymbol{c})$ and $\boldsymbol{a} \times (\boldsymbol{b} \times \boldsymbol{c})$.

Solution:

$$\boldsymbol{b} \times \boldsymbol{c} = \begin{vmatrix} \boldsymbol{i} & \boldsymbol{j} & \boldsymbol{k} \\ 2 & -1 & 2 \\ 1 & 1 & 0 \end{vmatrix}$$

$$= \boldsymbol{i}\begin{vmatrix} -1 & 2 \\ 1 & 0 \end{vmatrix} - \boldsymbol{j}\begin{vmatrix} 2 & 2 \\ 1 & 0 \end{vmatrix} + \boldsymbol{k}\begin{vmatrix} 2 & -1 \\ 1 & 1 \end{vmatrix} = -2\boldsymbol{i} + 2\boldsymbol{j} + 3\boldsymbol{k}.$$

Scalar triple product

This is used later for finding if the intersection of three planes has a unique solution. It is also used for finding the volume of a parallelepiped.

So:

$\boldsymbol{a}.(\boldsymbol{b} \times \boldsymbol{c}) = (\boldsymbol{i} + 2\boldsymbol{j} - \boldsymbol{k}).(-2\boldsymbol{i} + 2\boldsymbol{j} + 3\boldsymbol{k}) = -2 + 4 - 3 = -1$.

Vector triple product

$$\boldsymbol{a} \times (\boldsymbol{b} \times \boldsymbol{c}) = \begin{vmatrix} \boldsymbol{i} & \boldsymbol{j} & \boldsymbol{k} \\ 1 & 2 & -1 \\ -2 & 2 & 3 \end{vmatrix}$$

$$= \boldsymbol{i}\begin{vmatrix} 2 & -1 \\ 2 & 3 \end{vmatrix} - \boldsymbol{j}\begin{vmatrix} 1 & -1 \\ -2 & 3 \end{vmatrix} + \boldsymbol{k}\begin{vmatrix} 1 & 2 \\ -2 & 2 \end{vmatrix} = 8\boldsymbol{i} - \boldsymbol{j} + 6\boldsymbol{k}.$$

Although you do not **need** this, it is worth noting the vector product identity:
$A \times (B \times C) = B(A.C) - C(A.B)$, known as the BAC to CAB rule.

 ONLINE

Find out more about the scalar triple product at www.brightredbooks.net

 ONLINE

For more on the cross product and vector product applications, head to www.brightredbooks.net

ONLINE TEST

Test yourself on vectors at www.brightredbooks.net

THINGS TO DO AND THINK ABOUT

1 Given $\boldsymbol{u} = \boldsymbol{i} + \boldsymbol{j} - \boldsymbol{k}$, $\boldsymbol{v} = 2\boldsymbol{i} - \boldsymbol{j} + 2\boldsymbol{k}$ and $\boldsymbol{w} = -\boldsymbol{i} + 2\boldsymbol{k}$, calculate $\boldsymbol{u}.(\boldsymbol{v} \times \boldsymbol{w})$.   4

2 $\boldsymbol{a} = \begin{bmatrix} 2 \\ -1 \\ 3 \end{bmatrix}$ $\boldsymbol{b} = \begin{bmatrix} -1 \\ 0 \\ 1 \end{bmatrix}$ $\boldsymbol{c} = \begin{bmatrix} -2 \\ t \\ -1 \end{bmatrix}$

Find the value of t for which $\boldsymbol{c}$ is perpendicular to $\boldsymbol{a} \times \boldsymbol{b}$.   4

EQUATIONS OF LINES

EQUATION OF A LINE

A line L in space is determined by its direction vector, $\boldsymbol{d}$, and any point Q on the line.

For any point P on L, we have $QP = t\boldsymbol{d}$ for some scalar t, and as t varies from $-\infty$ to $+\infty$ we get all points on L.

If A is a point on L with position vector $\boldsymbol{a}$, and P has position vector $\boldsymbol{r}$, then $\boldsymbol{r} = \boldsymbol{a} + t\boldsymbol{d}$. This is the **vector equation** of a line.

If $\boldsymbol{r} = x\boldsymbol{i} + y\boldsymbol{j} + z\boldsymbol{k}$, and $\boldsymbol{q} = a\boldsymbol{i} + b\boldsymbol{j} + c\boldsymbol{k}$, and $\boldsymbol{d} = d\boldsymbol{i} + d\boldsymbol{j} + d\boldsymbol{k}$, then equating the $\boldsymbol{i}, \boldsymbol{j}$ and $\boldsymbol{k}$ components gives the **parametric equations** of a line:

$$x = a + td_1 \qquad y = b + td_2 \qquad z = c + td_3.$$

An equivalent form to these equations are the **symmetric** (or **Cartesian**) **equations**:

$$\frac{x-a}{d_1} = \frac{y-b}{d_2} = \frac{z-c}{d_3} = t$$

Although the equation is often not given in exam questions with the parameter shown, you should always include it in your working.

DON'T FORGET

There are three forms for the equation of a line:

Vector: $\boldsymbol{r} = \boldsymbol{a} + t\boldsymbol{d}$

Parametric: $x = a + td_1$
$\qquad\qquad y = b + td_2$
$\qquad\qquad z = c + td_3$

Symmetric:
$\frac{x-a}{d_1} = \frac{y-b}{d_2} = \frac{z-c}{d_3} = t$

Example: 9.4

Obtain all three forms (vector, parametric and symmetric) for the equation of the line which passes through the points A $(1, -1, 3)$ and B $(3, 0, 1)$.

Solution:

We need the direction vector $\boldsymbol{d}$.

This is: $\quad \overrightarrow{AB} = \begin{bmatrix} 3 \\ 0 \\ 1 \end{bmatrix} - \begin{bmatrix} 1 \\ -1 \\ 3 \end{bmatrix} = \begin{bmatrix} 2 \\ 1 \\ -2 \end{bmatrix}$

so: $\boldsymbol{d} = 2\boldsymbol{i} + \boldsymbol{j} - 2\boldsymbol{k}$.

The **vector equation** is given by:

$x\boldsymbol{i} + y\boldsymbol{j} + z\boldsymbol{k} = \boldsymbol{i} - \boldsymbol{j} + 3\boldsymbol{k} + t(2\boldsymbol{i} + \boldsymbol{j} - 2\boldsymbol{k})$. (Either point, A or B, may be used to get the equations.)

The **parametric equations** using A are:

$x = 1 + 2t, \quad y = -1 + t, \quad z = 3 - 2t$.

The **symmetric form** is given by:

$\frac{x-1}{2} = \frac{y+1}{1} = \frac{z-3}{-2} = t$.

Note that, if we use B, we get the parametric equations:

$x = 3 + 2\mu, \quad y = \mu, \quad z = 1 - 2\mu$.

Although these look quite different, they describe the same line. For example, $\mu = -1$ gives the point A.

At Higher:
For the equation of a line in two dimensions, you needed:
- a gradient and a point, or
- two points.

At Advanced Higher:
For the equation of a line in three dimensions, you need:
- a vector and a point, or
- two points.

DISTANCE OF A POINT FROM A LINE

Given a point P and a line L, the distance of P from L is the length PQ, where Q is the point on L such that $\overrightarrow{PQ}$ is perpendicular to the direction of L.

VIDEO LINK

Watch the clip at www.brightredbooks.net to see the distance of a point to a line in three dimensions using three different techniques.

Example: 9.5

Obtain the distance of the point P (1, 1, 1) from the line L with parametric equations:

$x = 2 + t, \quad y = 2 - t, \quad z = 1 + 2t.$

Solution:

Q has coordinates $(2 + t, 2 - t, 1 + 2t)$ for a value of t to be determined.

Hence $\quad \overrightarrow{PQ} = \begin{bmatrix} 2 + t \\ 2 - t \\ 1 + 2t \end{bmatrix} - \begin{bmatrix} 1 \\ 1 \\ 1 \end{bmatrix} = \begin{bmatrix} 1 + t \\ 1 - t \\ 2t \end{bmatrix}$

The direction vector of L is $\mathbf{d} = \begin{bmatrix} 1 \\ -1 \\ 2 \end{bmatrix}$, and we require $\overrightarrow{PQ}.\mathbf{d} = 0$.

This gives $(1 + t) - (1 - t) + 4t = 0$, so $t = 0$.

So, Q has coordinates $(2, 2, 1)$ and:

$PQ = \sqrt{(2 - 1)^2 + (2 - 1)^2 + (1 - 1)^2} = \sqrt{2}$.

So, the distance of P from L is $\sqrt{2}$.

VIDEO LINK

Check out the clip at www.brightredbooks.net for more on the equation of a line.

THINGS TO DO AND THINK ABOUT

Lines L_1 and L_2 are given by the equations

$L_1: \dfrac{x + 3}{-2} = \dfrac{y - 1}{1} = \dfrac{z - 5}{3} \qquad L_2: \dfrac{x - 1}{1} = \dfrac{y - 1}{-1} = \dfrac{z - 3}{-1}.$

(a) Show that L_1 and L_2 do not intersect. **3**

(b) Obtain parametric equations for the line L_3 which passes through P (–1, 0, 2) and has direction vector perpendicular to the directions of L_1 and L_2. **3**

(c) Show that L_3 intersects L_2, and obtain the point of intersection Q. **3**

(d) Verify that P lies on L_1, and explain why the distance PQ gives the shortest distance between L_1 and L_2. **1**

ONLINE TEST

Test yourself on this topic at www.brightredbooks.net

EQUATIONS OF PLANES 1 GPS

THE CARTESIAN EQUATION OF THE PLANE

If R is a general point on the plane with position vector r, and A is a given point on the plane with position vector a, then we can obtain the vector $\overrightarrow{AR}$ which will lie in the plane. $\overrightarrow{AR}$ will be perpendicular to the vector perpendicular to the plane, the normal vector, n.

Thus the vector $\overrightarrow{AR}$ will lie in the plane and will be perpendicular to n.

$$\overrightarrow{AR} \cdot n = 0$$
$$(r - a) \times n = 0$$
$$r \times n = a \times n$$

DON'T FORGET ✚

$r \times n = a \times n$

With a and n known, the scalar product $a \times n$ can be evaluated.

If $a \times n = d$, then the equation of the plane is $r \times n = d$.

This is the scalar product form of the vector equation of the plane.

If $n = ai + bj + ck$ and $r = xi + yj + zk$, then the Cartesian form of the equation of the plane is
$ax + by + cz = d$

If the plane passes through the origin, then $d = 0$.

Example: 9.6

Obtain the Cartesian equation of the plane passing through the points
$P\ (1, 1, 2)$, $Q\ (1, -1, 1)$ and $R\ (2, 4, 2)$.

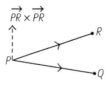

Solution:

Method 1

The vector $\overrightarrow{PQ} \times \overrightarrow{PR}$ is normal to the plane.

$$\overrightarrow{PQ} = \overrightarrow{OQ} - \overrightarrow{OP} = \begin{bmatrix} 1 \\ -1 \\ 1 \end{bmatrix} - \begin{bmatrix} 1 \\ 1 \\ 2 \end{bmatrix} = \begin{bmatrix} 0 \\ -2 \\ -1 \end{bmatrix} \qquad \overrightarrow{PR} = \overrightarrow{OR} - \overrightarrow{OP} = \begin{bmatrix} 2 \\ 4 \\ 2 \end{bmatrix} - \begin{bmatrix} 1 \\ 1 \\ 2 \end{bmatrix} = \begin{bmatrix} 1 \\ 3 \\ 0 \end{bmatrix}$$

$$\overrightarrow{PQ} \times \overrightarrow{PR} = \begin{vmatrix} i & j & k \\ 0 & -2 & -1 \\ 1 & 3 & 0 \end{vmatrix} = 3i - j + 2k.$$

So, a vector normal to the plane is:

$$3i - j + 2k, \quad n = \begin{bmatrix} 3 \\ -1 \\ 2 \end{bmatrix}$$

and the equation of the plane is of the form $3x - y + 2z = d$.

Substituting the coordinates of P: $(3(1) - (1) + 2(2) = 6)$ gives $d = 6$. As a check, use the coordinates of Q or R to give this same d value.

So, the equation of the plane is $3x - y + 2z = 6$.

contd

Method 2

The equation of the plane has the form $ax + by + cz = d$, so substituting the coordinates for P, Q and R gives the equations:

$a + b + 2c = d$, $a - b + c = d$, $2a + 4b + 2c = d$.

Solving these for a, b and c using the methods of pp 40–43 gives:

$a = \frac{d}{2}$, $b = -\frac{d}{6}$ and $c = \frac{d}{3}$

Substituting into the equation $ax + by + cz = d$, and simplifying, gives the equation $3x - y + 2z = 6$.

OBTAINING AN EQUATION OF A PLANE

To obtain an equation of a plane, we need one of the following:
- three points on the plane

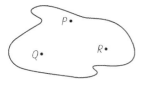

- two non-parallel lines lying in the plane

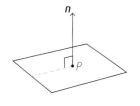

- a point on the plane and a direction vector normal to the plane

 VIDEO LINK

Learn more about equations of planes by watching the clips at www.brightredbooks.net

THINGS TO DO AND THINK ABOUT

Find the equation of the plane through $A(1, 2, -1)$, $B(2, 3, 1)$ and $C(1, 4, -2)$ in Cartesian form.

 ONLINE TEST

Test yourself on equations of planes at www.brightredbooks.net

EQUATIONS OF PLANES 2 GPS

THE VECTOR EQUATION OF THE PLANE

If R is a general point on the plane with position vector r, and A is a given point on the plane with position vector a, then $r = a + sb + tc$ ($s, t \in \mathbb{R}$) is the vector equation of the plane, where b and c are non-parallel vectors lying in the plane.

If we are given three (non-collinear) points A, B and C with position vectors a, b and c, then the plane is uniquely defined.

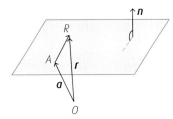

Consider a general point R in the plane having position vector r.

We can express the vector $\overrightarrow{AR}$ as a linear combination of two vectors in the plane:

$\overrightarrow{AR} = \lambda \overrightarrow{AB} + \mu \overrightarrow{AC}$

$r - a = \lambda(b - a) + \mu(c - a)$.

Rearranging, we get the vector equation of the plane, given three position vectors:

$r = (1 - \lambda - \mu)a + \lambda b + \mu c$.

When given a point on the plane and two vectors parallel to the plane or in-plane (direction vectors), the vector equation of the plane is given as

$r = a + \lambda d_1 + \mu d_2$.

When two lines in the plane are given, then the direction vectors of each line and a point on either line would be used.

Example: 9.7

Obtain an equation for the plane containing the lines:

$L_1: x = 1 + 2t, \qquad y = 2 - t, \qquad z = 1 - 2t$

$L_2: x = 1 - m, \qquad y = 2 + m, \qquad z = 1 + 3m$.

contd

Solution:

We first check that the lines intersect: setting the x-coordinates of the two lines equal, and also the y-coordinates, gives

$1 + 2t = 1 - m$ and $2 - t = 2 + m,$ so $t = m = 0.$

These values also make the z-coordinates equal, so the two lines intersect at the point P (1, 2, 1).

This gives us a point on the plane, and we have the direction vectors of the lines which lie in the plane, (2 −1 −2) and (−1 1 3),

so the **vector equation** is given by $r = \begin{bmatrix} 1 \\ 2 \\ 1 \end{bmatrix} + \lambda \begin{bmatrix} 2 \\ -1 \\ -2 \end{bmatrix} + \mu \begin{bmatrix} -1 \\ 1 \\ 3 \end{bmatrix}.$

To find the **Cartesian equation**, we find the cross product of the direction vectors of the lines which lie in the plane.

The direction of the normal to the plane containing L_1 and L_2 is given by:

$$\begin{vmatrix} i & j & k \\ 2 & -1 & -2 \\ -1 & 1 & 3 \end{vmatrix} = i \begin{vmatrix} -1 & -2 \\ 1 & 3 \end{vmatrix} - j \begin{vmatrix} 2 & -2 \\ -1 & 3 \end{vmatrix} + k \begin{vmatrix} 2 & -1 \\ -1 & 1 \end{vmatrix} = -i - 4j + k$$

so the equation of the plane has the form:

$-x - 4y + z = d.$

Substituting the intersection point $x = 1$, $y = 2$, $z = 1$ gives $d = -8.$

So (dividing through by −1), an equation for the plane is:

$x + 4y - z = 8.$

From the vector form $r = a + sb + tc$, if $a = a_1 i + a_2 j + a_3 k$, $b = b_1 i + b_2 j + b_3 k$ and $c = c_1 i + c_2 j + c_3 k$, then the equation of the plane can be written in the following **parametric form** where $r = x i + y j + z k$.

$x = a_1 + sb_1 + tc_1$

$y = a_2 + sb_2 + tc_2$

$z = a_3 + sb_3 + tc_3$

Any form of the equation of a plane may be given unless a particular one is asked for.

INTERSECTION OF TWO PLANES

Unless the planes are parallel, two planes intersect in a line.

Example: 9.8

Obtain parametric equations for the line of intersection of the planes with equations:

$x + 2y - z = 1$ and $2x - y + 2z = 2.$

Solution:

Let $z = t$, where t is a parameter, and solve the equations for x and y:

$x + 2y = 1 + t,$ $2x - y = 2 - 2t.$

Solving for x and y gives:

$x = 1 - \frac{3}{5}t,$ $y = \frac{4}{5}t,$ $z = t.$

As t is a parameter, let $t = 5s$ to get rid of fractions.

This gives the parametric equations for the line of intersection as:

$x = 1 - 3s,$ $y = 4s,$ $z = 5s.$

VIDEO LINK

See another example of finding the line of intersection of two planes at www.brightredbooks.net

VIDEO LINK

What happens when the planes are parallel? See this and further examples at www.brightredbooks.net

ONLINE TEST

Test yourself on equations of planes at www.brightredbooks.net

THINGS TO DO AND THINK ABOUT

Find the equation of the plane through A(1, 2, −1), B(2, 3, 1) and C(1, 4, −2) in vector form.

INTERSECTIONS OF PLANES GPS

VIDEO LINK

Do the three planes meet, or are they coplanar? Check this out at www.brightredbooks.net

VIDEO LINK

These planes are not coplanar, and they meet at a unique point (using the triple scalar product). Check it out at www.brightredbooks.net

VIDEO LINK

The clip at www.brightredbooks.net continues the story of how three planes can exist in three-dimensional space and how to find their line of intersection.

DON'T FORGET

There are several cases that can arise when three planes intersect:
- unique point of intersection
- line of intersection
- no intersection.

INTERSECTION OF THREE PLANES

Unique point of intersection

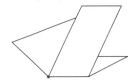

The unique point of intersection corresponds to the diagram shown here. Two of the planes intersect in a line, which intersects the third plane at a unique point. It corresponds to the situation where three simultaneous equations have a unique solution.

Example: 9.9

Obtain the point of intersection of the three planes given by:

$$x + y - 2z = 1, \quad x + 2y + z = 6, \quad 2x - y - z = -1.$$

Solution:

Using the Gaussian elimination methods given in pp 40–43, the (unique) solution is found to be:

$$x = 1, \quad y = 2, \quad z = 1.$$

Hence the three planes all meet at the point with coordinates (1, 2, 1).

Line of intersection

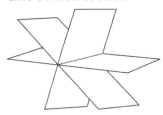

The line of intersection corresponds to the diagram shown here. The third plane contains the line of intersection of the other two planes. It corresponds to the situation where three simultaneous equations have infinitely many solutions.

Example: 9.10

Show that the planes with equations:

$$x + y + 2z = 1 \quad x - y + z = 2 \qquad 3x + y + 5z = 4$$

intersect in a line, giving parametric equations for this line.

Solution:

Using the Gaussian elimination methods in Chapter 5, we get:

$$
\begin{array}{ccc|c}
1 & 1 & 2 & 1 \\
1 & -1 & 1 & 2 \\
3 & 1 & 5 & 4
\end{array}
\quad \text{becomes:} \quad
\begin{array}{ccc|cl}
1 & 1 & 2 & 1 & \\
0 & -2 & -1 & 1 & R_2 - R_1 \\
0 & -2 & -1 & 1 & R_3 - 3R_1
\end{array}
$$

Now the last row becomes: $\quad 0 \quad 0 \quad 0 \mid \quad 0$

so, we set $z = t$, where t is a parameter.

Solving for y, then x, we get:

$$x = \tfrac{3}{2} - \tfrac{3}{2}t, \quad y = -\tfrac{1}{2} - \tfrac{1}{2}t, \quad z = t.$$

These are parametric equations of a line through the point $\left(\tfrac{3}{2}, -\tfrac{1}{2}, 0\right)$ with direction vector

parallel to $\begin{bmatrix} 3 \\ 1 \\ -2 \end{bmatrix}$ so the three planes meet in a line.

contd

No intersection

The situation when there is no intersection between three planes corresponds to one of the diagrams shown here.

A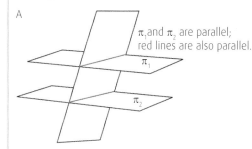

π_1 and π_2 are parallel; red lines are also parallel.

B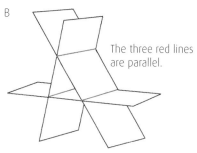

The three red lines are parallel.

It corresponds to the case where the simultaneous equations are inconsistent.

Example: 9.11

Show that the planes with equations:

$x + y + z = 1, \quad 2x + y - z = 2, \quad 4x + 3y + z = 1$

do not meet at a single point. Obtain equations for the lines of intersection:

$$
\begin{array}{ccc|c}
1 & 1 & 1 & 1 \\
2 & 1 & -1 & 2 \\
4 & 3 & 1 & 1
\end{array}
\quad \text{becomes:} \quad
\begin{array}{ccc|cl}
1 & 1 & 1 & 1 & \\
0 & -1 & -3 & 0 & R_2 - 2R_1 \\
0 & -1 & -3 & -3 & R_3 - 4R_1
\end{array}
$$

Solution:

Now $R_3 - R_2$ gives the last row as: $\quad 0 \quad 0 \quad 0 \mid \quad -3$

so the three equations are inconsistent.

This means the three planes do not meet at a single point.

Setting $z = t$ in all three equations and solving them in pairs gives equations of the lines of intersection of the planes taken in pairs. The details are left as an exercise, giving:

- planes 1 and 2:   line $x = 1 + 2t,$ $y = -3t,$ $z = t$
- planes 1 and 3:   line $x = -2 + 2t,$ $y = 3 - 3t,$ $z = t$
- planes 2 and 3:   line $x = \frac{5}{2} + 2t,$ $y = -3 - 3t,$ $z = t$

This example illustrates the right-hand diagram (B) above.

 ## THINGS TO DO AND THINK ABOUT

Intersection of three planes, in a nutshell:

Let n_1, n_2 and n_3 represent the normal vectors of three distinct planes.

If $n_1 . n_2 \times n_3 \neq 0$, then

- normal vectors are not coplanar
- there is a single point of intersection.

If $n_1 . n_2 \times n_3 = 0$, then

- the normal vectors are coplanar
- there may or may not be points of intersection
- if points of intersection exist, then it will be a line of intersection.

$n_1 . n_2 \times n_3$ or $n_2 . n_1 \times n_3$ or $n_3 . n_1 \times n_2$ may be used.

 VIDEO LINK

See another example, where the line is given as a parametric equation, explained at www.brightredbooks.net

 VIDEO LINK

For a vector equation of the intersecting line, see the clip at www.brightredbooks.net

 ONLINE TEST

Test yourself on vectors at www.brightredbooks.net

INTERSECTIONS OF LINES AND PLANES GPS

INTERSECTION OF TWO LINES

When solving this type of problem, it is essential to use different parameters for the two different lines.

Example: 9.12

Lines L_1 and L_2 are defined as follows:

L_1: $x = 3 + t$, $y = 4 + 2t$, $z = -t$

L_2: $x = 3 + 2m$, $y = 3 + 3m$, $z = 4 + 2m$.

Decide if they intersect – and, if they do, obtain the point of intersection.

DON'T FORGET

Use different parameters for two different lines.

Solution:

The lines intersect if the following set of three equations in two unknowns has a solution:

$3 + t = 3 + 2m$, $4 + 2t = 3 + 3m$, $-t = 4 + 2m$.

These equations are formed by equating x from L_1 with x from L_2 etc.

The procedure is to solve any pair of equations and **check that this solution satisfies the third equation**. If it does, the value of t or m will give the point of intersection. If it doesn't, the lines do not intersect.

Solving the first two equations gives $t = -2$ and $m = -1$, and these satisfy the third equation, so the lines intersect.

Putting $t = -2$ in the equations for L_1 (or $m = -1$ in L_2) gives the intersection point $(1, 0, 2)$.

DON'T FORGET

When checking if two lines meet, all three equations for the parameters must be checked.

INTERSECTION OF A LINE AND PLANE

The best way to solve this type of problem is to use the standard equation of the plane and a parametric equation of the line.

Example: 9.13

Obtain the point of intersection of the line given by $x = 2 - t$, $y = -2t$ $z = 1 + 2t$ and the plane with equation $x + 2y + z = 0$.

Solution:

All we need to do is substitute the parametric equations of the line into the equation of the plane, so:

$(2 - t) + 2(-2t) + (1 + 2t) = 0$.

Solving gives $t = 1$, and the point of intersection is given by setting $t = 1$ in the parametric equations of the line. This gives the point of intersection as $(1, -2, 3)$.

Two other situations can arise:

- the line is parallel to the plane
- the line lies completely in the plane.

Example: 9.14

Show that, for all values of a except one, the line given by:

$x = a + t, y = 2 + t, z = 1 - t$

does not intersect the plane with the equation: $2x - y + z = 1$.

What happens for this exceptional value of a?

contd

Solution:

Substituting the parametric equations into the equation of the plane gives:

$2(a + t) - (2 + t) + (1 - t) = 1 \Rightarrow 2a - 1 = 1$.

Note that the t terms cancel out, and for all $a \neq 1$ we are left with an impossibility. Hence, for all values of a except 1, the line does not intersect the plane (they are parallel).

When $a = 1$, we get the identity $1 = 1$. This means that the coordinates of every point on the line satisfy the condition $2x - y + z = 1$ automatically, i.e. the line lies completely in the plane.

DISTANCE OF A POINT FROM A PLANE

Given a point P and a plane π, the distance of P from π is the length PQ, where Q is the point on π such that $\overrightarrow{PQ}$ is perpendicular to π.

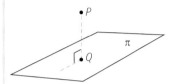

VIDEO LINK

Learn more about this topic by watching the clip at www.brightredbooks.net

Example: 9.15

Obtain the distance of the point P (2, –1, 3) from the plane π with equation:

$2x + y - z = 6$.

Solution:

Let Q be the point on π such that $\overrightarrow{PQ}$ is perpendicular to π, which means that the direction vector of the line PQ is $2\boldsymbol{i} + \boldsymbol{j} - \boldsymbol{k}$ (here the direction vector of the line is obviously the same as the normal vector to the plane).

Hence the parametric equations of the line PQ are:

$x = 2 + 2t, \quad y = -1 + t, \quad z = 3 - t$.

This line meets π when $2(2 + 2t) + (-1 + t) - (3 - t) = 6$, so $t = 1$.

Putting $t = 1$ in the parametric equations gives Q (4, 0, 2).

The distance of P from π is thus $\sqrt{(2 - 4)^2 + (-1 - 0)^2 + (3 - 2)^2} = \sqrt{6}$.

ONLINE TEST

Test yourself on the intersections of lines and planes at www.brightredbooks.net

THINGS TO DO AND THINK ABOUT

1   (a) Obtain an equation for the plane passing through the point A (2, 1, –1) which is perpendicular to the line L given by $\frac{x + 3}{1} = \frac{y}{2} = \frac{z - 12}{1}$.　　3

    (b) Obtain the coordinates of the point B where the plane and L intersect.　　4

    (c) Calculate the distance AB, and explain why this gives the shortest distance from A to L.　　3

ANGLES BETWEEN LINES AND PLANES

Given two directions determined by vectors $\boldsymbol{a}$ and $\boldsymbol{b}$, we have $\boldsymbol{a}.\boldsymbol{b} = |\boldsymbol{a}||\boldsymbol{b}| \cos \theta$.

So $\cos \theta = \frac{\boldsymbol{a}.\boldsymbol{b}}{|\boldsymbol{a}||\boldsymbol{b}|}$, and this gives the angle between two vectors.

 DON'T FORGET

$\cos \theta = \frac{\boldsymbol{a}.\boldsymbol{b}}{|\boldsymbol{a}||\boldsymbol{b}|}$

ANGLE BETWEEN TWO LINES

Remember that lines in space may not intersect, so it only makes sense to calculate the angle between intersecting lines.

Example: 9.16

Given that they intersect, calculate the acute angle between the lines:

L_1: $x = 1 + t$, $y = 2t$, $z = 2 - t$
L_2: $x = 3 + 2s$, $y = 3 + 3s$, $z = 4 + 2s$.

Solution:

Direction vectors for the lines are:

$\boldsymbol{d}_1 = \boldsymbol{i} + 2\boldsymbol{j} - \boldsymbol{k}$ and $\boldsymbol{d}_2 = 2\boldsymbol{i} + 3\boldsymbol{j} + 2\boldsymbol{k}$

$|\boldsymbol{d}_1| = \sqrt{1 + 4 + 1} = \sqrt{6}$ and $|\boldsymbol{d}_2| = \sqrt{4 + 9 + 4} = \sqrt{17}$

$\boldsymbol{d}_1.\boldsymbol{d}_2 = 2 + 6 - 2 = 6$.

The angle between the lines is given by:

$\cos \theta = \frac{6}{\sqrt{6}\sqrt{17}} = \frac{\sqrt{6}}{\sqrt{17}}$, so:

$\theta = \cos^{-1}\left(\frac{\sqrt{6}}{\sqrt{17}}\right) = 53.6°$.

Unless told otherwise, 3 significant figures is sufficient.

ANGLE BETWEEN A LINE AND A PLANE

Assume that the line is not parallel to the plane. Because different lines in the plane will make different angles with this intersecting line, we calculate the acute angle $\theta°$ between the intersecting line and the normal to the plane, and the angle between the line and the plane is $(90 - \theta)°$.

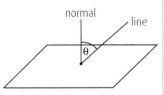

 ONLINE

Head to www.brightredbooks.net to explore this topic further.

Example: 9.17

Calculate the acute angle between the line given by:

$x = 2 - t$, $y = -2t$, $z = 1 + 2t$

and the plane with equation:

$x + y + z = 0$.

Solution:

A direction vector for the line is $\boldsymbol{d} = \begin{bmatrix} -1 \\ -2 \\ 2 \end{bmatrix}$, and a normal vector for the plane is $\boldsymbol{n} = \begin{bmatrix} 1 \\ 1 \\ 1 \end{bmatrix}$,

so: $\cos \theta = \frac{-1 - 2 + 2}{\sqrt{1 + 1 + 1} \times \sqrt{1 + 4 + 4}} = \frac{-1}{3\sqrt{3}}$.

Hence $\theta = 101.1°$.

The acute angle is $180° - 101.1° = 78.9°$, which is the angle between the line and the normal to the plane, so the angle between the line and plane is $90° - 78.9° = 11.1°$.

ANGLE BETWEEN TWO PLANES

The angle between two planes is the angle between their normal vectors.

Example: 9.18

For what value of α do the planes with equations

$x + 2y - z = 1$ and $\alpha x - y + 3z = 2$

intersect at right angles?

Solution:

If the angle between the planes, θ, is 90°, then $\cos\theta = 0$.

The normals are $\mathbf{n}_1 = \begin{bmatrix} 1 \\ 2 \\ -1 \end{bmatrix}$ and $\mathbf{n}_2 = \begin{bmatrix} \alpha \\ -1 \\ 3 \end{bmatrix}$, and we need $\mathbf{n}_1 . \mathbf{n}_2 = 0$.

So, we require $1 \times \alpha + 2 \times (-1) + (-1) \times 3 = 0$. This gives $\alpha = 5$.

 **DON'T FORGET**

When calculating the angle between vectors $\mathbf{a}$ and $\mathbf{b}$, always start with $\mathbf{a}.\mathbf{b}$. If this is zero, then the angle between them is 90° without further calculation.

 VIDEO LINK

See another example online at www.brightredbooks.net

THINGS TO DO AND THINK ABOUT

1 (a) Use Gaussian elimination to solve the system of equations
$2x - y + z = 8$, $x + 2y + z = 3$, $-x + 3y + 2z = 1$.   5

  (b) Show that the line of intersection, L, of the planes $2x - y + z = 8$ and $x + 2y + z = 3$ has parametric equations
$x = 5 - 3t$, $y = -t$, $z = 5t - 2$.   2

  (c) Calculate the acute angle between the line L and the plane $-x + 3y + 2z = 1$.   4

2 (a) Find an equation for the plane, π_1, containing the points:
$A\,(3, 4, 4)$, $B\,(-1, -8, 4)$ and $C\,(2, 5, 6)$.   4

  (b) Given that $\pi_2 = 3x + y + 3z = 25$ and $\pi_3 = 3x - y - z = -5$, determine the nature of the intersections, if any, of the planes π_1, π_2 and π_3.   4

  (c) Calculate the size of the acute angle between planes π_2 and π_3.   2

3 (a) Find an equation for the plane, π_1, that contains the points:
$A\,(1, 2, 0)$, $B\,(-1, 6, 4)$ and $C\,(2, -3, -2)$.   4

  (b) Find the acute angle between π_1 and the line
$\frac{a+2}{3} = \frac{y-1}{2} = \frac{z+1}{-1}$ 3

 **ONLINE**

See this mathematics applied in the geometry of crystals online at www.brightredbooks.net

 **ONLINE TEST**

Test yourself on this topic at www.brightredbooks.net

DIFFERENTIAL EQUATIONS

FIRST-ORDER EQUATIONS 1 MAC

ONLINE

Find a practical example of using a differential equation at www.brightredbooks.net

OVERVIEW

An equation for an unknown function y of x which involves derivatives of y is called a **differential equation**. The highest derivative in the equation determines its order.

The equation $\frac{dy}{dx} = x + y$ is a first-order equation.

The equation $\frac{d^2y}{dx^2} + \frac{dy}{dx} - y = \sin x$ is a second-order equation.

DON'T FORGET

$f(y)dy = g(x)dx \Rightarrow$
$\int f(y)dy = \int g(x)dx$

VARIABLES SEPARABLE

The easiest equations to deal with are called **separable**. In these equations, we can move all the y-terms to one side of the equation and all the x-terms to the other side.

First-order separable equations are equations which can be put in the form:

$f(y)\frac{dy}{dx} = g(x)$, where f and g are given functions.

So:

$\int f(y)dy = \int g(x)dx$.

Integrating both sides gives: $F(y) = G(x) + c$

where c is an arbitrary constant. This is the **general solution**.

The general solution of a first-order differential equation contains one arbitrary constant.

From this, we can usually express y as a function of x, although sometimes the solution has to be left in implicit form.

If we are given the value of y for a given value of x (this is called an **initial condition**), we can obtain the value of the constant c.

Example: 10.1

Obtain the general solution of the differential equation $sec^2\, y\frac{dy}{dx} = 2x$.

Solution:

Separating the variables, we have: $sec^2\, y\, dy = 2x\, dx$

$\int sec^2\, y\, dy = \int 2x dx$

$\tan y = x^2 + c$, so: $y = \tan^{-1}(x^2 + c)$.

 DON'T FORGET

Remember the arbitrary constant 'c'.

Example: 10.2

Obtain the general solution of the differential equation $\frac{dy}{dx} = \frac{x}{y+1}$, and the particular solution for which $y = 1$ when $x = 0$.

Solution:

Separating the variables, we have: $(y + 1)dy = xdx$,

so: $\int(y + 1)dy = \int xdx \Rightarrow \frac{(y+1)^2}{2} = \frac{x^2}{2} + c$.

This can be written as:

$(y + 1)^2 = x^2 + k$, where $k\ (= 2c)$ is the arbitrary constant.

Solving for y gives: $y = -1 \pm\sqrt{x^2 + k}$.

There are two families of solutions: $y = -1 +\sqrt{x^2 + k}$ and $y = -1 -\sqrt{x^2 + k}$.

For the particular solution $y = 1$ when $x = 0$,

$1 = -1 +\sqrt{0^2 + k}$ or $1 = -1 -\sqrt{0^2 + k}$

$\sqrt{k} = 1 + 1 = 2$ or   not possible: $\sqrt{0^2 + k} \neq -2$

$\Rightarrow k = 4$

So $y = -1 +\sqrt{x^2 + 4}$.

DON'T FORGET

Add c as soon as you integrate before any attempt to solve for y in terms of x.

 contd

Laws of growth and decay

An important application of separable first-order differential equations arises in simple models of growth and decay.

Example: 10.3

A circular fungus grows by absorbing nutrients at its boundary. The radius of the fungus r is increasing at the rate $10 - r$ mm per day. At time $t = 0$, the fungus has radius 1mm.

Obtain r as a function of t. Does the fungus grow without limit?

Solution:

$\frac{dr}{dt} = 10 - r$, so: $\frac{dr}{10 - r} = dt \qquad \Rightarrow \quad -\ln|10 - r| = t + c.$

When $t = 0$, $r = 1$, so: $\quad c = -\ln 9 \quad \Rightarrow \quad -\ln|10 - r| = t - \ln 9.$

Hence: $\ln 9 - \ln|10 - r| = t \qquad \Rightarrow \quad \ln\left|\frac{9}{10 - r}\right| = t$

and solving for r gives: $\quad r = 10 - 9e^{-t}.$

From this, we can see that $r < 10$ for all t, and so the fungus will never have a radius greater than 10 mm.

If a population has unlimited resources to enable it to grow, the rate at which the population grows is proportional to the size of the population. So, if the size is N at time t, we have:

$\frac{dN}{dt} = kN$, where k is a positive constant.

k is a positive arbitrary constant, so N increases exponentially.

A model of population growth which has an 'overcrowding' factor is:

$\frac{dN}{dt} = kN(N_0 - N)$

where k and N_0 are known positive constants. To solve this differential equation, we need partial fractions.

Changes in world population growth

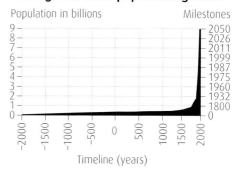

Population in billions   Milestones

Timeline (years)

Example: 10.4

Solve the differential equation $\frac{dN}{dt} = N(100 - N)$ given that $N(0) = 50$ and $N < 100$.

What happens to N as $t \to +\infty$?

Solution:

$\frac{dN}{N(100 - N)} = dt \quad \Rightarrow \quad \int\frac{dN}{N(100 - N)} = t + c.$

Using partial fractions: $\quad \frac{1}{N(100 - N)} = \frac{1}{100}\left(\frac{1}{N} + \frac{1}{100 - N}\right)$, so:

$\int\frac{dN}{N(100 - N)} = \frac{1}{100}\int\left(\frac{1}{N} + \frac{1}{100 - N}\right)dN = \frac{1}{100}[\ln N - \ln|100 - N|]$

$= \frac{1}{100}\ln\left|\frac{N}{100 - N}\right|.$

Hence: $\frac{1}{100}\ln\left|\frac{N}{100 - N}\right| = t + C$, and $N = 50$ when $t = 0$ gives $C = 0$.

So: $\ln\left|\frac{N}{100 - N}\right| = 100t \quad \Rightarrow \quad \frac{N}{100 - N} = e^{100t}$ (we have $N < 100$).

Solving for N gives $N = e^{100t}(100 - N)$, so: $N + Ne^{100t} = 100e^{100t}.$

Hence $N = \frac{100e^{100t}}{1 + e^{100t}} = \frac{100}{1 + e^{-100t}} \quad \Rightarrow \quad N \to 100$ as $t \to +\infty.$

 ## THINGS TO DO AND THINK ABOUT

A young man carrying a contagious virus returns to an isolated village with just 1000 inhabitants. Determine the number of infected people after six days if it is observed that 50 people are affected after four days. You may assume that the rate of infection is described by the differential equation

$\frac{dx}{dt} = kx(1000 - x)$, $x(0) = 1.$

 ## ONLINE TEST

Test yourself on this topic at www.brightredbooks.net

6

FIRST-ORDER EQUATIONS 2 MAC

INTEGRATING FACTORS

Differential equations of the following type cannot be solved by the previous method:

$$\frac{dy}{dx} + a(x)y = b(x).$$

By multiplying by a suitable function of x, the left-hand side can be made into an exact derivative, and then both sides of the equation can be integrated. This function is called the **integrating factor** (IF).

The integrating factor for $\frac{dy}{dx} + a(x)y = b(x)$ is $e^{\int a(x)dx}$.

Note that: $p(x)\frac{dy}{dx} + q(x)y = r(x)$

may be written as: $\frac{dy}{dx} + \frac{q(x)}{p(x)}y = \frac{r(x)}{p(x)}$

so: $a(x) = \frac{q(x)}{p(x)}$ and $b(x) = \frac{r(x)}{p(x)}$, giving $\frac{dy}{dx} + a(x)y = b(x)$

The differential equation above, in which the coefficient of $\frac{dy}{dx}$ is 1, is called the **standard form**.

This is called a **linear differential equation** because there are no powers of y or products of y with $\frac{dy}{dx}$.

Steps for solving equations using the integrating factor

1   Write in standard form

$$\frac{dy}{dx} + a(x)y = b(x)$$

2   Obtain *IF* (the integrating factor) where $IF = e^{\int a(x)dx}$

3   Multiply the standard form by *IF* to get:

$$IF\frac{dy}{dx} + IFa(x)y = IFb(x)$$

4   $\frac{d}{dx}(IF.y) = IF.b(x)$

5   $IF.y = \int IF.b(x)dx$

6   $y = \frac{\int IF.b(x)dx}{IF}$

Example: 10.5

Obtain the general solution of the equation $\frac{dy}{dx} + \frac{2}{x}y = \frac{\sin x}{x^2}$ $(x > 0)$.

Solution:

$\int \frac{2}{x}dx = 2\ln x = \ln(x^2)$, so the integrating factor is: $e^{\ln(x^2)} = x^2$.

So, $x^2\frac{dy}{dx} + 2xy = \sin x$, i.e. $\frac{d}{dx}(x^2y) = \sin x \Rightarrow x^2y = -\cos x + C$,

and so the general solution is: $y = -\frac{\cos x}{x^2} + \frac{C}{x^2}$.

contd

Example: 10.6

Obtain the solution of $\sin x \frac{dy}{dx} - 2y \cos x = 3\sin^3 x$ for which $y = 0$ when $x = \frac{\pi}{2}$.

Solution:

We first need to divide through by $\sin x$ to put the equation in the correct form:

$\frac{dy}{dx} - \frac{2\cos x}{\sin x}y = 3\sin^2 x$, so the integrating factor is:

$\exp\left(\int \frac{-2\cos x}{\sin x}dx\right) = \exp(-2\ln \sin x) = \exp(\ln(\sin x)^{-2}) = \frac{1}{\sin^2 x}$.

Don't forget the minus sign.

Multiplying by the integrating factor gives: $\frac{1}{\sin^2 x}\frac{dy}{dx} - \frac{2\cos x}{\sin^3 x}y = 3$,

so: $\frac{d}{dx}\left(\frac{y}{\sin^2 x}\right) = 3 \Rightarrow \frac{y}{\sin^2 x} = 3x + C$

and the general solution is:

$y = (3x + C)\sin^2 x$.

As $y = 0$ when $x = \frac{\pi}{2}$, $C = -\frac{3\pi}{2}$, giving the solution $y = \left(3x - \frac{3\pi}{2}\right)\sin^2 x$.

 **DON'T FORGET**

The constant C must be introduced when you integrate $b(x)e^{A(x)}$.

🧠 THINGS TO DO AND THINK ABOUT

 VIDEO LINK

For more on this topic, check out the clip at www.brightredbooks.net

1 Solve the differential equation

$x\frac{dy}{dx} - 2y = x^3 e^{2x}$

given that $y(1) = 0$. Express your answer in the form $y = f(x)$.   6

2 The production manager in a chemical factory models a process with the differential equation:

$x^2\frac{dy}{dx} - 3xy + 1 = 0$.

(a) Given $y = 2$ when $x = \frac{1}{2}$, find the particular solution, giving y in terms of x.   6

 ONLINE TEST

Test yourself on differential equations at www.brightredbooks.net

Latest production methods suggested a better model would be:

$\frac{1}{y}\frac{dy}{dx} - 3x = 1$.

(b) Given $y = 1$ when $x = 0$, find the particular solution to the new differential equation model.   4

(c) Calculate the percentage difference in yield, y, for the new process when $x = 1$.   2

3 The population, P, of a particular insect, when introduced into a small, controlled environment, can be modelled by

$\frac{dP}{dt} = \frac{16k - P}{1000}$

where t is the time in hours after the initial population was introduced.

(a) Show that the general solution of the differential equation can be expressed as

$P = 16k - \frac{1}{Ae^{\frac{t}{1000}}}$ 4

(b) Given that the initial population of 100 had grown to 200 after 1000 hours, find the exact values of A and k.   4

(c) Hence state the long-term maximum population.   2

SECOND-ORDER EQUATIONS 1 MAC

Differential equations have a remarkable ability to predict the world around us. They are used in a wide variety of disciplines, from biology to economics, physics, chemistry and engineering. They can describe exponential growth and decay, the population growth of species or the change in investment return over time.

DON'T FORGET

The general solution of second-order differential equations requires two arbitrary constants.

HOMOGENEOUS SECOND-ORDER DIFFERENTIAL EQUATIONS

The equation $a\frac{d^2y}{dx^2} + b\frac{dy}{dx} + cy = 0$ is called a **homogeneous equation** because the right-hand side is zero.

The general solution of this equation depends on the quadratic equation $am^2 + bm + c = 0$. This is called the **auxiliary equation**. There are three separate cases to consider, depending on the nature of the roots. These three cases are:

- real unequal roots
- equal roots
- complex roots.

General solutions for these three cases are shown below. Note that A and B are arbitrary constants.

1 Real unequal roots α, β. The general solution is: $y = Ae^{\alpha x} + Be^{\beta x}$.

2 Equal roots α, α. The general solution is: $y = Ae^{\alpha x} + Bxe^{\alpha x}$ (or $y = (A + Bx)\,e^{\alpha x}$).

3 Complex roots $p \pm qi$. The general solution is: $y = e^{px}(A \cos qx + B \sin qx)$.

Example: 10.7

Obtain the general solution of the equation $\frac{d^2y}{dx^2} + 4\frac{dy}{dx} - 5y = 0$.

Solution:

The auxiliary equation is:

$m^2 + 4m - 5 = 0$

$\Rightarrow (m - 1)(m + 5) = 0$.

The roots are 1 and −5, so the general solution is:

$y = Ae^x + Be^{-5x}$.

To evaluate two arbitrary constants, we need two pieces of information which come in the form of initial conditions, such as where the values of y and $\frac{dy}{dx}$ are given for a given value of x.

Example: 10.8

Obtain the solution of $\frac{d^2y}{dx^2} - 4\frac{dy}{dx} + 4y = 0$ for which $y = 2$ and $\frac{dy}{dx} = 1$ when $x = 0$.

contd

Solution:

The auxiliary equation is:

$m^2 - 4m + 4 = 0$

$(m - 2)^2 = 0$.

This gives roots = 2, 2.

The general solution is $y = (A + Bx)e^{2x}$

$\frac{dy}{dx} = Be^{2x} + 2(A + Bx)e^{2x}$.

Setting $y = 2$ when $x = 0$ gives $A = 2$.

$\frac{dy}{dx} = 1$ when $x = 0$ gives: $B + 2A = 1$, so $B = -3$.

The solution is: $y = (2 - 3x)e^{2x}$.

ONLINE

Head to
www.brightredbooks.
net for more on solving
differential equations.

Example: 10.9

Obtain the solution of

$\frac{d^2y}{dx^2} + 8\frac{dy}{dx} + 20y = 0$, given that $y = 1$ when $x = 0$ and $y = 2$ when $x = \frac{\pi}{4}$.

Solution:

The auxiliary equation is:

$m^2 + 8m + 20 = 0$.

By completing the square, $(m + 4)^2 = -4$, so:

$m = -4 \pm 2i$.

Alternatively, by using the quadratic formula:

$m = \frac{-8 \pm \sqrt{8^2 - 4.1.20}}{2} = \frac{-8 \pm 4i}{2} = -4 \pm 2i$.

Hence the general solution is:

$y = e^{-4x}(A \cos 2x + B \sin 2x)$.

Setting $y = 1$ when $x = 0$

gives $A = 1$, and $y = 2$

when $x = \frac{\pi}{4}$ gives $B = 2e^{\pi}$.

The solution is:

$y = e^{-4x}(\cos 2x + 2e^{\pi} \sin 2x)$.

THINGS TO DO AND THINK ABOUT

ONLINE TEST

Test yourself on this topic
at www.brightredbooks.net

1 Solve the differential equation
   $\frac{d^2y}{dx^2} + 6\frac{dy}{dx} + 9y = 0$
   given that $y = 2$ and $\frac{dy}{dx} = -3$ when $x = 0$.   7

2 Obtain the general solution of the differential equation
   $\frac{d^2y}{dx^2} + 4\frac{dy}{dx} - 5y = 0$.
   Obtain the particular solution for which $y(0) = 1$ and $y'(0) = 10$.   6

3 Solve the differential equation
   $\frac{d^2y}{dx^2} + 2\frac{dy}{dx} + 5y = 0$
   given that $y = 1$ when $x = 0$ and $y = 2$ when $x = \frac{\pi}{4}$.   6

SECOND-ORDER EQUATIONS 2 (MAC)

NON-HOMOGENEOUS SECOND-ORDER DIFFERENTIAL EQUATIONS

DON'T FORGET

General solution = CF + PI

To solve $\quad a\frac{d^2y}{dx^2} + b\frac{dy}{dx} + cy = f(x)$,

first solve the homogeneous equation $\quad a\frac{d^2y}{dx^2} + b\frac{dy}{dx} + cy = 0$.

This gives the **complementary function (CF)**.

We then find a **particular integral (PI)** (methods given below), which leads to the **general solution** of the non-homogeneous equation.

General solution = complementary function + particular integral.

DON'T FORGET

Any initial conditions must be applied to the general solution of the full equation (CF + PI).

Example: 10.10

Verify that $x + 1$ is a particular integral of the equation:
$$\frac{d^2y}{dx^2} - 3\frac{dy}{dx} + 2y = 2x - 1$$
and hence obtain the general solution of this equation.

Solution:

DON'T FORGET

$' = \frac{d}{dx}$

If $y = x + 1$, $\frac{dy}{dx} = 1$ and $\frac{d^2y}{dx^2} = 0$, so:
$$\frac{d^2y}{dx^2} - 3\frac{dy}{dx} + 2y = 0 - 3(1) + 2(x + 1) = 2x - 1.$$
The auxiliary equation for
$$\frac{d^2y}{dx^2} - 3\frac{dy}{dx} + 2y = 0 \text{ is:}$$
$m^2 - 3m + 2 = 0$, which has roots 1 and 2, and general solution:

$y = Ae^x + Be^{2x}$.

Hence the general solution of $\frac{d^2y}{dx^2} - 3\frac{dy}{dx} + 2y = 2x - 1$ is:

$y = Ae^x + Be^{2x} + x + 1$.

Methods for finding a particular integral

For certain types of function f, it is possible to guess the general form of the particular integral and then evaluate any unknown parameters. This is not the same as evaluating arbitrary constants in a general solution. The table below should be remembered.

Forms of $f(x)$	Forms of particular integral
Polynomials $3x + 4$ $6x^2 + 2$ $10x^2 + 3x$	 $ax + b$ $ax^2 + bx + c$ $ax^2 + bx + c$
Trigonometric functions $4\cos x$ $3\sin 2x$	 $a\cos x + b\sin x$ $a\cos 2x + b\sin 2x$
Exponential functions $2e^{3x}$	 ae^{3x}

Example: 10.11

Obtain the general solution of the equation $\frac{d^2y}{dx^2} + 3\frac{dy}{dx} - 4y = 4e^{-x}$, and hence obtain the particular solution satisfying $y = 2$ and $\frac{dy}{dx} = 3$ when $x = 0$.

contd

Solution:

Substituting into the differential equation:

$ae^{-x} - 3ae^{-x} - 4ae^{-x} = 4e^{-x} \Rightarrow -6ae^{-x} = 4e^{-x}$

so: $a = -\frac{2}{3}$.

The general equation of $\frac{d^2y}{dx^2} + 3\frac{dy}{dx} - 4y = 4e^{-x}$ is:

$y = CF + PI$ so: $y = Ae^x + Be^{-4x} - \frac{2}{3}e^{-x}$.

If $y = 2$ when $x = 0$, then $A + B = \frac{8}{3}$.

Also: $\frac{dy}{dx} = Ae^x - 4Be^{-4x} + \frac{2}{3}e^{-x}$, so:

$\frac{dy}{dx} = 3$ when $x = 0$ gives $A - 4B = \frac{7}{3}$.

Solving for A and B gives: $A = \frac{13}{5}$, $B = \frac{1}{15}$.

The solution satisfying $y = 2$ and $\frac{dy}{dx} = 3$ when $x = 0$ is:

$y = \frac{13}{5}e^x + \frac{1}{15}e^{-4x} - \frac{2}{3}e^{-x}$.

ONLINE

To read more about the use of differential equations, take a look at www.brightredbooks.net

Example: 10.12

Obtain the general solution of the equation:

$\frac{d^2y}{dx^2} + 6\frac{dy}{dx} + 25y = 2\sin x$.

VIDEO LINK

See solutions being worked out and speed revision at www.brightredbooks.net

Solution:

The auxiliary equation for the homogeneous equation is:

$m^2 + 6m + 25 = 0$.

Completing the square (or using the quadratic formula) gives:

$(m + 3)^2 = -16 \Rightarrow m = -3 \pm 4i$, so the general solution of the homogeneous equation (i.e. the complementary function) is: $y = e^{-3x}(A \cos 4x + B \sin 4x)$.

For a particular integral of the full equation, let $y = a \cos x + b \sin x$:

$\frac{dy}{dx} = -a \sin x + b \cos x$ and $\frac{d^2y}{dx^2} = -a \cos x - b \sin x$.

Substituting into the full equation gives:

$-a \cos x - b \sin x + 6(b \cos x - a \sin x) + 25(a \cos x + b \sin x) = 2\sin x$.

This is an identity, so, after collecting like terms on the left-hand side, we equate the coefficients of $\cos x$ and $\sin x$ on both sides:

$\cos x$: $-a + 6b + 25a = 0$

$\sin x$: $-b - 6a + 25b = 2$.

Solving the two equations $24a + 6b = 0$ and $-6a + 24b = 2$ simultaneously gives:

$a = -\frac{1}{51}$, $b = \frac{4}{51}$.

So the general solution of $\frac{d^2y}{dx^2} + 6\frac{dy}{dx} + 25y = 2\sin x$ is:

$y = e^{-3x}(A \cos 4x + B \sin 4x) + \frac{4}{51} \sin x - \frac{1}{51} \cos x$.

THINGS TO DO AND THINK ABOUT

1 Obtain the general solution of the differential equation
   $\frac{d^2y}{dx^2} + 4\frac{dy}{dx} - 5y = 7e^{2x} + 10$. **7**

   Obtain the particular solution for which $y(0) = 1$ and $y'(0) = 10$. **3**

2 Obtain the general solution of the differential equation
   $y'' + 4y' + 8y = 8x^2 + 16x + 6 \left(' = \frac{d}{dx}\right)$. **6**

ONLINE TEST

Test yourself on this topic at www.brightredbooks.net

NUMBER THEORY AND PROOF

NUMBERS, NOTATION AND DIRECT PROOF

Numbers are classified according to type, and you need to be familiar with natural numbers, integers, primes, rational and irrational numbers, real numbers and complex numbers.

NATURAL NUMBERS

Natural numbers are the counting numbers, 1, 2, 3, … . The set of natural numbers is often denoted by $\mathbb{N}$, so the notation to show that 2 is a natural number is: $2 \in \mathbb{N}$ (remembering that $\in$ means 'is a member of'). $\mathbb{N}_0$ denotes the natural numbers, but with 0 included. This is the set of whole numbers and is sometimes referred to as $\mathbb{W}$.

ONLINE

Learn more about prime numbers by following the link at www.brightredbooks.net

INTEGERS

When we add 0 and the negative numbers $-1, -2, …$ to the natural numbers, we get the set of **integers**, usually denoted by $\mathbb{Z}$. $\mathbb{N}$ is a subset of $\mathbb{Z}$. So, $-2 \in \mathbb{Z}$.

An important subset of $\mathbb{N}$, and hence also of $\mathbb{Z}$ and $\mathbb{R}$, is the set of **prime numbers**, $\mathbb{P}$. These are the positive integers which cannot be factorised. The first six are 2, 3, 5, 7, 11, 13.

The fundamental theorem of arithmetic

ONLINE

Discover more about the fundamental theorem of arithmetic by following the links at www.brightredbooks.net

The fundamental theorem of arithmetic states that every integer greater than 1 can be written as a product of primes in one, and only one, way.

Example: 11.1

Factorise 16 500.

Solution:

As the number is clearly even, start by dividing by 2 and continue for as long as possible:
$16\,500 \div 2 \div 2 = 4125$, which is odd, so we now try dividing by 3:

$4125 \div 3 = 1375$, but this is not divisible by 3, so next try 5:

$1375 \div 5 \div 5 \div 5 = 11$, and as 11 is prime, we stop.

$16\,500 = 2^2 \times 3 \times 5^3 \times 11$.

ONLINE

Head to www.brightredbooks.net for more on irrational numbers.

IRRATIONAL AND RATIONAL NUMBERS

Rational numbers are of the form $\frac{m}{n}$ ($n \neq 0$), where m and n are integers.

This set is denoted by $\mathbb{Q}$. $\mathbb{Z}$ is a subset of $\mathbb{Q}$.

Irrational numbers cannot be expressed in the form $\frac{m}{n}$. An example of an irrational number is $\sqrt{2}$.

REAL NUMBERS

Real numbers are either rational or irrational. The set of all real numbers is usually denoted by $\mathbb{R}$. $\mathbb{Q}$ is a subset of $\mathbb{R}$.

COMPLEX NUMBERS

Complex numbers are of the form $x + iy$, where x and y are real numbers, and $i = \sqrt{-1}$.
The set of complex numbers is usually denoted by $\mathbb{C}$. $\mathbb{R}$ is a subset of $\mathbb{C}$.

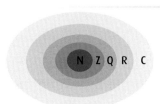

NOTATION

There are symbols used in mathematical proofs which, when used properly, make the proofs clearer and more concise. Here are symbols that it would be useful to know and be able to use:

$\neg P$ 'not P' or 'statement P is false'. Also written as ~P.

$\therefore$ 'Therefore'

$P \Rightarrow Q$ 'P true implies Q true' or 'if P is true, then Q is also true'.

$P \Leftrightarrow Q$ '$P \Rightarrow Q$ and $Q \Rightarrow P$' or 'P true if and only if Q true'.

$\in$ 'belongs to' or 'belonging to', for example $2 \in \mathbb{N}$ means 2 belongs to $\mathbb{N}$.

$\forall$ 'for all', for example $\forall n \in \mathbb{N}$ means 'for all natural numbers'.

$\exists$ 'there exists'.

$\nexists$ 'there does not exist'.

$\neq$ 'not equal to'.

$|$ 'divides', for example $4|n$ means n is divisible by 4. This assumes $n \in \mathbb{N}$.

$\equiv$ 'is equivalent to' or 'is congruent to', so true for all possible values of variables. For example, $x^2 - y^2 \equiv (x + y)(x - y)$, $x, y \in \mathbb{R}$.
In geometry, two or more figures are identical: e.g. $\triangle ABC \equiv \triangle DEF \equiv \triangle GHI$ means that the three triangles ABC, DEF and GHI are congruent.

DIRECT PROOF

The simplest and most straightforward type of proof is direct proof.

Example: 11.2

This is a simple *direct proof* that the sum of two even integers is itself an even number.

Consider two even integers x and y.

Since they are even, they can be written as $x = 2a$ and $y = 2b$ respectively for integers a and b. Then the sum $x + y = 2a + 2b = 2(a + b)$.

From this, it is clear that $x + y$ has 2 as a factor and therefore is even, so the sum of any two even integers is even.

Example: 11.3

Prove that the sum of any two rational numbers is rational.

Solution:

This proof follows directly from the definition of what it means for a number to be rational. Given that r and s are rational numbers, show that $r + s$ is rational.

Proof: Since r and s are rational, we can write $r = \frac{p}{q}$ and $s = \frac{m}{n}$ for some integers p, q, m and n. Then $r + s = \frac{p}{q} + \frac{m}{n} = \frac{pn + qm}{qn}$.

Since p, q, m and n are integers, $pn + qm$ and qn are also integers (the sum or product of integers is an integer). Thus, by the calculation above, $r + s$ is the quotient of two integers, and is therefore a rational number.

Here's an example of a proof that is really just a calculation.

Example: 11.4

Given the trigonometric identity $\sin(x + y) = \sin x \cos y + \cos x \sin y$, prove the identity: $\sin(2x) = 2 \sin x \cos x$.

Proof: $\sin(2x) = \sin(x + x) = \sin x \cos x + \cos x \sin x$ (by the sum identity)
          $= 2 \sin x \cos x$.

Note that this proof is merely a string of equalities connecting $\sin(2x)$ to $2 \sin x \cos x$. Many proofs are like this when proving an identity.

THINGS TO DO AND THINK ABOUT

1   Prove that $\sec^2 x = 1 + \tan^2 x$.   3

2   Prove that the product of two positive integers, where one is odd and the other even, is even.   2

DON'T FORGET

Work from one side towards the other.

DON'T FORGET

You cannot use what you are proving as part of your proof!

DON'T FORGET

Be familiar with, and know how to use, all trig formulae you have met previously.

ONLINE TEST

Test yourself on numbers, notation and direct proof at www.brightredbooks.net

PROOF BY CONTRADICTION AND BY CONTRAPOSITIVE

PROOF BY CONTRADICTION

Suppose we have a statement, A, and we want to prove that A is true. One way of doing this is to assume that A is false, but that this leads to a contradiction. This is the method of **proof by contradiction**.

A common question in assessments is to prove that, for some non-square integer a, $\sqrt{a}$ is irrational.

The first method below holds for all such a.

The second method is shorter for the special case where a is even.

Method 1

Example: 11.5

Prove by contradiction that $\sqrt{7}$ is irrational.

Solution:

Since $\sqrt{7}$ must be either rational or irrational, assume that it is rational.

Then $\sqrt{7} = \frac{m}{n}$ for integers m and n, which have no common factors, apart from 1, because we can cancel any common factor first $[(m, n) = 1]$.

$$\sqrt{7} = \frac{m}{n}$$
$$7 = \frac{m^2}{n^2}$$
$$7n^2 = m^2$$

$\therefore\ 7|m$ by the fundamental theorem of arithmetic

$\therefore\ 7^2|m^2$

$\therefore\ 7^2|7n^2$

$\therefore\ 7|n^2$

$\therefore\ 7|n$ by the fundamental theorem of arithmetic

$\therefore\ 7|m$ and $7|n$, i.e. m and n have a common factor.

This contradicts $(m, n) = 1$.

This in turn contradicts our original assumption that $\sqrt{7}$ is rational, and hence $\sqrt{7}$ must be irrational.

Method 2

Example: 11.6

Prove by contradiction that $\sqrt{6}$ is irrational.

Solution:

Since $\sqrt{6}$ must be either rational or irrational, assume that it is rational.

Then $\sqrt{6} = \frac{m}{n}$ for integers m and n which have no common factors, apart from 1, because we can cancel any common factor first $[(m, n) = 1]$.

So: $6 = \frac{m^2}{n^2} \Rightarrow m^2 = 6n^2$
$$\Rightarrow m^2 = 2 \times 3n^2$$

and hence m^2 is even. It follows that m is even, $m = 2r$ $(r \in \mathbb{Z})$.

> This can be assumed from the fundamental theorem of arithmetic. For a proof of this property, see example 11.6 below.

So, $4r^2 = m^2 = 6n^2 \Rightarrow 3n^2 = 2r^2$, and hence $3n^2$ is even. But 3 is odd, so n^2 must be even, and hence n must be even.

This means that both m and n are even, which is a contradiction because we started with m and n having no common factor. So, the assumption that $\sqrt{6}$ is rational has led to a contradiction, and hence $\sqrt{6}$ must be irrational.

DON'T FORGET

$(m, n) = 1$ means the greatest common divisor of m and n is 1.

DON'T FORGET

An even integer can be written as $2m$ and an odd integer as $2m - 1$ or $2m + 1$, where m is an integer.

ONLINE

Head to www.brightredbooks.net to learn more about contradiction.

PROOF BY CONTRAPOSITIVE

A useful technique where a direct proof may be impractical or cumbersome is proof by contrapositive. Suppose we have two statements, P and Q, and wish to prove that $P \Rightarrow Q$. This is equivalent to proving that $\neg Q \Rightarrow \neg P$, that is, Q false $\Rightarrow P$ false. This is **proof by contrapositive**. To illustrate this, consider a class of 10 girls and 10 boys. Three of the boys are colour-blind, but none of the girls is. Then it is true that:

    (i) A pupil is colour-blind $\Rightarrow$ That pupil is a boy.

and

    (ii) A pupil is not a boy $\Rightarrow$ That pupil is not colour-blind.

(i) and **(ii)** are equivalent.

Example: 11.7

Consider these two statements:

A: n^2 is even

B: n is even ($n \in \mathbb{Z}$)

Prove that $A \Rightarrow B$, using the contrapositive.

Solution:

Assume that B is false. Then n is odd and $n = 2m + 1$ where $m \in \mathbb{Z}$.

$$(2m + 1)^2 = 4m^2 + 4m + 1$$
$$= 2(2m^2 + 2m) + 1$$
$$= 2r + 1$$

where $r \in \mathbb{Z}$.

This shows that $(2m + 1)^2 = n^2$ is odd, so not even, i.e. $\sim A$.

So, we have $\sim B \Rightarrow \sim A$, $\therefore$ by the contrapositive if n^2 is even, then n is even.

Example: 11.8

Consider the statements:

P: $x^2 - 6x + 5$ is even

Q: x is odd ($x \in \mathbb{Z}$)

Prove that $P \Rightarrow Q$, using the contrapositive.

Solution:

Consider $\neg Q$, that is x is not odd and therefore is even.

$x = 2m$, $m \in \mathbb{N}$

$$x^2 - 6x + 5 = (2m)^2 - 6(2m) + 5$$
$$= 4m^2 - 12m + 5$$
$$= 4m^2 - 12m + 4 + 1$$
$$= 2(2m^2 - 6m + 2) + 1$$
$$= 2n + 1,\ n \in \mathbb{Z},\ \text{since } m \in \mathbb{Z}$$

$\therefore x^2 - 6x + 5$ is odd, i.e. $\neg Q \Rightarrow \neg P$.

Therefore, by the contrapositive, $P \Rightarrow Q$.

Thus, if $x^2 - 6x + 5$ is even, then x is odd.

DON'T FORGET

'~' or '¬' may be used for **not**.

 THINGS TO DO AND THINK ABOUT

In this question, the method of proof is given. In the Advanced Higher paper, the method may not be specified. You should choose the most appropriate method of proof, although alternatives could be acceptable for full marks.

1 Prove, by use of the contrapositive, that if x is an irrational number, then $3 + x$ is irrational.

    4

 ONLINE TEST

Test yourself on proof by contradiction and contrapositive at www.brightredbooks.net

FURTHER PROOF

COUNTEREXAMPLES

To prove a mathematical statement, it is not enough to list numerical examples which support it, however many you give – because just one example is enough to disprove a statement. Such an example is called a **counterexample**. This is a useful method to disprove a statement which claims to be true 'for all possible values of n, $\forall n$'.

Example: 11.9

Prove or disprove the statement: 'If a function f with domain $\mathbb{R}$ satisfies $f''(0) = 0$, then f has a point of inflexion at $x = 0$'.

Solution:

The statement is false. A counterexample is the function $f(x) = x^4$ (because f'' doesn't change sign at $x = 0$, the curve remains concave up).

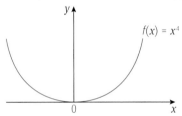

$f''(x) = 12x^2 \geqslant 0$, so always concave up; $f''(0) = 0$, but $x = 0$ is not a point of inflexion.

Example: 11.10

Consider the statement: 'If the product of two integers m and n is even, both integers are even'. Either prove this result or give a counterexample.

Solution:

The statement isn't true; one counterexample is $m = 2$, $n = 3$.

ONLINE

You can try out your knowledge at www.brightredbooks.net

NECESSARY AND SUFFICIENT

Consider two statements A and B. If $A \Rightarrow B$, we say that A is sufficient for B and that B is necessary for A (because if B isn't true, A can't be either).

Sometimes we have both $A \Rightarrow B$ and $B \Rightarrow A$. In this case, we write $A \Leftrightarrow B$ and say that A is necessary and sufficient for B.

Example: 11.11

Consider these statements A and B about integers m and n:

A m and n are odd

B mn is odd.

Decide whether A is (a) sufficient, (b) necessary, (c) necessary and sufficient for B.

contd

Solution:

(a) Does $A \Rightarrow B$?

Let $m = 2p + 1$, $n = 2q + 1$ (p, q integers).

($2p + 1$ and $2p - 1$ would not be general enough; we need distinct p, q).

Then $mn = (2p + 1)(2q + 1) = 4pq + 2p + 2q + 1 = 2(2pq + p + q) + 1$, which is odd. Hence A is sufficient for B.

(b) Does $B \Rightarrow A$? We prove this is true by contradiction. Assume that mn is odd, and suppose at least one of m, n is even. If $m = 2p$ then $mn = 2pn$, which is even, contradicting the assumption that mn is odd. Hence $B \Rightarrow A$, and A is necessary for B.

(c) Because both $A \Rightarrow B$ and $B \Rightarrow A$, we have $A \Leftrightarrow B$.

Hence A is necessary and sufficient for B.

DON'T FORGET

A sufficient for B means $A \Rightarrow B$.
A necessary for B means $B \Rightarrow A$.

IF AND ONLY IF

Another way of formulating 'necessary and sufficient' statements is to use 'if and only if'.

Example: 11.12

Prove that a polynomial $p(x) = a + bx + cx^2$ is an even function if and only if $b = 0$.

DON'T FORGET

These statements are all equivalent:
$A \Leftrightarrow B$
A is necessary and sufficient for B.
A is true if and only if B is true.

Solution:

There are two parts to an 'if and only if' proof:

'if' part:

If $b = 0$, then $p(x) = a + cx^2$,

so, $p(-x) = a + c(-x)^2 = a + cx^2 = p(x)$

$p(-x) = p(x) \Rightarrow p$ is an even function.

'only if' part:

Let the function p be even, so that $p(x) = p(-x)$ for all x.

Then $a + bx + cx^2 = a - bx + cx^2 \Rightarrow bx = -bx$ for all x, which gives $b = -b$, and so $b = 0$.

DON'T FORGET

An 'if and only if' proof usually consists of two parts.

 THINGS TO DO AND THINK ABOUT

 ONLINE TEST

Head to www.brightredbooks.net and test yourself on further proofs.

1 For each of the following statements, decide if it is true or false. Justify your conclusions.

  A The cube of any even integer p plus the square of any odd integer q is odd.

  B For all natural numbers m, if m^2 is divisible by 9, then m is divisible by 9.   5

2 Consider the number abc in base 10 (i.e. the integer $100a + 10b + c$).
Prove that abc is divisible by 9 if, and only if, $a + b + c$ is divisible by 9.   5

3 For each of statements A and B, if it is true, prove it, and if it is false, disprove it.

  A $p^3 - p$ is divisible by 6 for all integers $p > 2$.

  B $p^4 - p$ is divisible by 6 for all integers $p > 2$.   4

4 For each of statements A and B, if it is true, prove it, and if it is false, disprove it.

  A If n^3 is even, so is n.

  B If n^3 is a multiple of 8, so is n.   5

PROOF BY INDUCTION

Proof by induction is often used to prove results involving sums of terms. However, it also useful in other cases where only integer values for n are required.

Example: 11.13

Prove by induction that $4 + 7 + 10 + \ldots + (3n + 1) = \frac{1}{2}n(3n + 5)$ for all integers $n \geqslant 1$.

This can also be written using Σ notation as:

$4 + 7 + 10 + \ldots + (3n + 1) = \sum\limits_{r=1}^{n} (3r + 1)$.

Solution:

Step 1: Check that the formula holds for $n = 1$:

LHS = 4   RHS = $\frac{1}{2}(1)(3 + 5) = 4$

$\therefore$ true for $n = 1$.

Step 2: Assume that the formula has been proved for some integer k (we know that $k = 1$ works), so that $4 + 7 + 10 + \ldots + (3k + 1) = \frac{1}{2}k(3k + 5)$.

We now need to show that, given this result, the result with k replaced by $k + 1$ holds as well, i.e.:

$4 + 7 + \ldots + (3k + 1) + (3(k + 1) + 1) = \frac{1}{2}(k + 1)(3(k + 1) + 5)$

$4 + 7 + \ldots + (3k + 1) + (3k + 4) = \frac{1}{2}(k + 1)(3k + 8)$.

Step 3: We now have to do some algebra.

Because we are assuming the result for k, namely:

$4 + 7 + 10 + \ldots + (3k + 1) = \frac{1}{2}k(3k + 5)$

it follows that:

$4 + 7 + \ldots + (3k + 1) + (3k + 4) = \frac{1}{2}k(3k + 5) + (3k + 4)$.

We now have to show that the right-hand side of the equation is indeed $\frac{1}{2}(k + 1)(3k + 8)$.

Start with this RHS and take out a common factor:

$\frac{1}{2}k(3k + 5) + (3k + 4)$

$= \frac{1}{2}[k(3k + 5) + (6k + 8)]$

$= \frac{1}{2}(3k^2 + 11k + 8)$

$= \frac{1}{2}(3k + 8)(k + 1)$

$\therefore$ if true for $n = k$, then true for $n = k + 1$.

Step 4: $\therefore$ if true for $n = k$, then also true for $n = k + 1$.

Since also true for $n = 1$ then, by induction, true for all $n \geqslant 1$.

The induction method can also be used for proofs involving inequalities or divisibility statements.

Example: 11.14

Prove by induction that $n^2 > 3n + 1$ for all integers $n \geqslant 4$.

contd

Solution:

Step 1: True for $n = 4$ because $16 > 13$. (Note that the result isn't true for $n = 1, 2, 3$.)

Step 2: Assume that $k^2 > 3k + 1$ for some integer k. We need to deduce from this that $(k + 1)^2 > 3(k + 1) + 1 = 3k + 4$.

Step 3: This is where we do the algebra:

$(k + 1)^2 = k^2 + 2k + 1 > (3k + 1) + 2k + 1$ because we are assuming $k^2 > 3k + 1$.

So: $(k + 1)^2 > (3k + 1) + 2k + 1 = (3k + 4) + (2k - 2)$

and because $k > 1$ (actually $k \geq 4$): $(3k + 4) + (2k - 2) > 3k + 4$.

Hence we have shown that $k^2 > 3k + 1$ implies that $(k + 1)^2 > 3(k + 1) + 1$.

Step 4: $\therefore$ if true for $n = k$, then also true for $n = k + 1$.
Since also true for $n = 4$, then, by induction, true for all $n \geq 4$.

 **DON'T FORGET**

Take care that the wording of the final statement is appropriate.

 ONLINE

Follow the links at www.brightredbooks.net for more on this topic.

Example: 11.15

Prove by induction that $6^n + 4$ is divisible by 10 for all integers $n \geq 1$.

Solution:

Step 1: True for $n = 1$, because $6 + 4 = 10$. $10 \,|\, 10$

Step 2: Assume for some k that $6^k + 4$ is divisible by 10, so that:

$6^k + 4 = 10m$ (or $6^k = 10m - 4$) for some integer m.

Step 3: Consider $6^{k+1} + 4$: $6^{k+1} + 4 = 6 \times 6^k + 4$
$= 6(10m - 4) + 4$
$= 60m - 20$
$= 10(6m - 2)$.

So, $6^k + 4$ divisible by 10 $\Rightarrow$ $6^{k+1} + 4$ divisible by 10.

Step 4: $\therefore$ if true for $n = k$, then also true for $n = k + 1$.
Since also true for $n = 1$ then, by induction, true for all $n \geq 1$.

 DON'T FORGET

Although a direct proof here may be easier, the question specifies induction as the method, so other methods would be penalised severely.

 **THINGS TO DO AND THINK ABOUT**

 **ONLINE TEST**

Test yourself on proof by induction at www.brightredbooks.net

1 Prove by induction that $\sum_{r=1}^{n}(6r^2 + 4r) = n(n + 1)(2n + 3)$.   5

2 Prove by induction on n that, for $x > 0$,

$(1 + x)^n \geq 1 + nx + \frac{1}{2}n(n - 1)x^2$

for all positive integers n.   5

3 Given that A and B are non-singular square matrices of same size, express the inverse of AB in terms of A^{-1} and B^{-1}.
Prove by induction on n that the inverse of A^n is $(A^{-1})^n$ for all integers $n \geq 1$.   5

4 Prove by induction that $\frac{d^n}{dx^n}(xe^x) = (x + n)e^x$ for all integers $n \geq 1$.   5

5 Prove that by induction that

$\sum_{r=1}^{n} r(r + 1) = \frac{1}{3}n(n + 1)(n + 2)$ 5

6 Prove by induction that, for all integers $n > 2$, $7^n - 4^{n-2}$ is divisible by 3.   6

7 Given that $f(x) = e^{px}$, prove by induction that the nth derivative of $f(x)$, denoted $f^{(n)}(x)$, is $p^n e^{px}$, for all positive integers, n.   5

8 Given the matrix $A = \begin{bmatrix} p & 0 \\ 0 & q \end{bmatrix}$, prove by induction that $A^n = \begin{bmatrix} p^n & 0 \\ 0 & q^n \end{bmatrix}$, for all $n \in \mathbb{N}$.   4

9 The nth integral of $f(x)$ is denoted $F^{(n)}(x)$. Given $f(x) = 1 + 2x$, prove by induction that

$F^{(n)}_{(x)} = \frac{(1 + 2x)^{n+1}}{(n + 1)!2^n}$ for all positive integers, n.   4

EUCLIDEAN ALGORITHM

For any positive integers a and b ($b \neq 0$), there exist unique integers q and r, where $0 \leqslant r < b$, such that: $a = bq + r$.

This is the **division algorithm**.

Example: 11.16

If $a = 1127$ and $b = 17$, find q and r ($0 \leqslant r < 17$) such that $1127 = 17q + r$.

Solution:

$\frac{1127}{17} = 66.29$, so consider $1127 - 66 \times 17$.

This gives $1127 = 66 \times 17 + 5$.

So $q = 66$, $r = 5$.

DON'T FORGET

The **greatest common divisor** (gcd) of two integers a and b is the largest integer that divides both a and b exactly.

When the gcd of a and b is 1, we say that they are **co-prime**.

The notation (a, b) for the gcd of a and b is often used.

Euclid's algorithm states that: if $a = bq + r$ where $0 \leqslant r < b$, then $(a, b) = (b, r)$.

Example: 11.17

Obtain the gcd of 1800 and 210.

Solution:

$1800 = 210 \times 8 + 120$ so $(1800, 210) = (210, 120)$

$210 = 120 \times 1 + 90$ so $(210, 120) = (120, 90)$

$120 = 90 \times 1 + 30$ so $(120, 90) = (90, 30)$

$90 = 30 \times 3 + 0$ so $(90, 30) = (30, 0) = 30$.

Hence the gcd of 1800 and 210 is 30. The statements in red do not need to be included in a solution to an examination question, but you may find them helpful.

Example: 11.18

Find integers s and t such that $1800s + 210t = 30$.

Solution:

From example 11.16, we have:

$30 = 120 - 90 \times 1$

$30 = 120 - (210 - 120 \times 1)$

$30 = 2 \times 120 - 210$

$30 = 2 \times (1800 - 8 \times 210) - 210$

$30 = 2 \times 1800 - 17 \times 210$

$\therefore s = 2, t = -17$.

DON'T FORGET

Be systematic about your layout.

The working for the two examples above may be combined as follows:

$1800 = 210 \times 8 + 120$	$\therefore 120 = 1800 - 210 \times 8$	[1]
$210 = 120 \times 1 + 90$	$\therefore 90 = 210 - 120$	[2]
$120 = 90 \times 1 + 30$	$\therefore 30 = 120 - 90$	[3]
$[2] \rightarrow [3]$	$30 = 120 - (210 - 120)$	
	$30 = 2 \times 120 - 210$	[4]
$[1] \rightarrow [4]$	$30 = 2(1800 - 210 \times 8) - 210$	
	$30 = 1800 \times 2 - 210 \times 17$	

$\therefore s = 2, t = -17$.

Example: 11.19

Find integers a and b such that $7425a + 2744b = 1$.

Solution:

We start by finding the gcd of 7425 and 2744.

$7425 = 2744 \times 2 + 1937$ so $(7425, 2744) = (2744, 1937)$
$2744 = 1937 \times 1 + 807$ so $(2744, 1937) = (1937, 807)$
$1937 = 807 \times 2 + 323$ so $(1937, 807) = (807, 323)$
$807 = 323 \times 2 + 161$ so $(807, 323) = (323, 161)$
$323 = 161 \times 2 + 1$ so $(323, 161) = (161, 1) = 1$.

This shows that the gcd of 7425 and 2744 is 1. As before, the statements in red can be omitted. Working backwards, we have:

$1 = 323 - 161 \times 2$
$ = 323 - 2 \times (807 - 323 \times 2)$
$ = 5 \times 323 - 2 \times 807$
$ = 5 \times (1937 - 807 \times 2) - 2 \times 807$
$ = 5 \times 1937 - 12 \times (2744 - 1937)$
$ = 17 \times 1937 - 12 \times 2744$
$ = 17 \times (7425 - 2744 \times 2) - 12 \times 2744$
$ = 17 \times 7425 - 46 \times 2744$.

So, $a = 17$ and $b = -46$.

DON'T FORGET

Take care with back substitution: numbers should get bigger.

CHANGING NUMBER BASES

The division algorithm can be used to change a number represented in one base to its representation in another base. A number m in base a can be written as m_a.

Example: 11.20

Express 315_6 as a number in base 3.

ONLINE

Follow the link at www.brightredbooks.net to explore this further.

Solution:

First, convert the number to base 10:

$$315_6 = \begin{array}{c|c|c} 36 & 6 & \text{U} \\ \hline 3 & 1 & 5 \end{array}$$

$315 = 3 \times 6^2 + 1 \times 6 + 5 \times 6^0 = 119_{10}$

We now use repeated division by 3:

$119 \div 3 = 39$ remainder 2
$39 \div 3 = 13$ remainder 0
$13 \div 3 = 4$ remainder 1
$4 \div 3 = 1$ remainder 1
$1 \div 3 = 0$ remainder 1

$119_{10} = 11\,102_3$

So, $315_6 = 11\,102_3$.

THINGS TO DO AND THINK ABOUT

1   Use the Euclidean algorithm to obtain the greatest common divisor of $29\,400$ and 6860, expressing it in the form $29\,400a + 6860b$, where a and b are integers.   **4**

2   Use the Euclidean algorithm to express 426_7 as a number in base 5.   **4**

3   Use the Euclidean algorithm to show that $(231, 17) = 1$, where (a, b) denotes the highest common factor of a and b.
   Hence find integers x and y such that $231x + 17y = 1$.   **4**

ONLINE TEST

Test yourself on the Euclidean algorithm at www.brightredbooks.net

APPENDICES

COURSE CONTENTS: SYLLABUS SKILLS CHECKLISTS

METHODS IN ALGEBRA AND CALCULUS

1.1 Applying algebraic skills to partial fractions

Express a proper rational fraction as a sum of partial fractions where the denominator may contain: distinct linear factors, an irreducible quadratic factor, a repeated factor	☐
Reduce an improper rational fraction to a polynomial and a proper rational fraction by division or otherwise	☐

1.2 Applying calculus through techniques of differentiation

Diffentiate functions involving e^x, ln x	☐
Apply the chain rule to differentiate the composition of at most three functions	☐
Differentiate functions of the form $f(x)g(x)$ and $\frac{f(x)}{g(x)}$	☐
Know the definitions of tan x and cot x	☐
Know the definitions of sec x and cosec x	☐
Derive and use derivatives of tan x, cot x, sec x, cosec x	☐
Differentiate functions which require more than one application or combination of applications of chain rule, product rule and quotient rule	☐
Know and use $\frac{dy}{dx} = \frac{1}{\frac{dx}{dy}}$	☐
Use logarithmic differentiation; recognise when it is appropriate in extended products and quotients and in functions where the variable occurs in an index	☐
Differentiate expressions of the form e.g. $\sin^{-1} kx$, $\tan^{-1} [f(x)]$	☐
Use differentiation to find the first derivative of a function defined implicitly, including in context	☐
Use differentiation to find the second derivative of a function defined implicitly	☐
Use differentiation to find the first derivative of a function defined parametrically, including in context	☐
Use differentiation to find the second derivative of a function defined parametrically	☐
Solve practical related rates by first establishing a functional relationship between appropriate variables	☐

1.3 Applying calculus skills through techniques of integration

Use $\int e^x \, dx$, $\int \frac{dx}{x}$, $\int \sec^2 x \, dx$	☐
Use the integrals of $\frac{1}{\sqrt{a^2 - x^2}}$, $\frac{1}{a^2 + x^2}$	☐
Recognise and integrate expressions of the form $\int g(f(x))f'(x)dx$ and $\int \frac{f'(x)}{f(x)} \, dx$	☐
Use partial fractions to integrate proper rational functions where the denominator may have: i) two separate or repeated linear factors ii) three linear factors with constant numerator iii) three linear factors with non-constant numerator iv) a linear factor and an irreducible quadratic factor of form $ax^2 + bx + c$	☐
Integrate where the substitution is given	☐

contd

Use integration by parts with one application	☐
Use integration by parts involving repeated applications	☐

1.4 Applying calculus skills to solving differential equations

Solve equations that can be written in the form $\frac{dy}{dx} = g(x)h(y)$ or $\frac{dy}{dx} = \frac{g(x)}{h(y)}$	☐
Find general and particular solutions given suitable information	☐
Solve equations by first writing linear equations in the standard form $\frac{dy}{dx} + P(x)y = f(x)$	☐
Find general and particular solutions given suitable information	☐
Find the general solution and particular solution of second-order homogeneous ordinary differential equations of the form $a\frac{d^2y}{dx^2} + b\frac{dy}{dx} + cy = 0$ with constant coefficients where the roots of the auxiliary equation are real or complex conjugates	☐
Solve second-order non-homogeneous ordinary differential equations of the form $a\frac{d^2y}{dx^2} + b\frac{dy}{dx} + cy = f(x)$ with constant coefficients using the auxiliary equation and particular integral method	☐

APPLICATIONS IN ALGEBRA AND CALCULUS

1.1 Applying algebraic skills to the binomial theorem and to complex numbers

Use the binomial theorem $(a + b)^n = \sum_{r=0}^{n} \binom{n}{r} a^{n-r} b^r$, for $r, n \in \mathbb{N}$	☐
Expand an expression of the form $(ax^p + by^q)^n$, where $a, b \in \mathbb{Q}$; $p, q \in \mathbb{Z}$; $n \leq 7$	☐
Using the general term for a binomial expansion, find a specific term in an expression	☐
Perform all of the operations of addition, subtraction, multiplication and division on complex numbers	☐
Find the square root of a complex number	☐
Find the roots of a cubic or quartic with real coefficients when one complex root is given	☐
Solve equations involving complex numbers	☐

1.2 Applying algebraic skills to sequences and series

Apply the rules on sequences and series to find the nth term, sum to n terms, sum to infinity, common difference and common ratio of arithmetic and geometric sequences respectively	☐
Determine the condition for a particular geometric series to converge	☐
Use the Maclaurin expansion to find a power series for a simple non-standard function	☐
Use the Maclaurin expansion to find a power series from two simpler series	☐

1.3 Applying algebraic skills to summation and mathematical proof

Know and use sums of certain series and other straightforward results and combinations thereof (formulae which appear on the formulae sheet are 'use' only)	☐
Use mathematical induction to prove summation formulae	☐
Use proof by induction	☐

1.4 Applying algebraic and calculus skills to properties of functions

Find the vertical asymptote to the graph of a rational function	☐
Find the non-vertical asymptote to the graph of a rational function	☐

contd

Investigate points of inflexion	☐
Investigate other features: stationary points, domain and range, symmetry (odd/even), continuous/discontinuous, extrema of functions; the maximum and minimum values of a continuous function f defined on a closed interval $[a, b]$ can occur at stationary points, end points or points where f' is not defined	☐
Sketch graphs using features given or obtained	☐
Sketch related functions: · modulus functions · inverse functions · differentiated functions · translations and reflections	☐

1.5 Applying algebraic and calculus skills to problems

Apply differentiation to problems in context	☐
Apply differentiation to optimisation	☐
Apply integration to volumes of revolution where the volume generated is by the rotation of the area under a single curve about the x- and y-axes	☐
Use calculus to determine corresponding connected integrals	☐
Apply integration to the evaluation of areas including integration with respect to y	☐

GEOMETRY, PROOF AND SYSTEMS OF EQUATIONS

1.1 Applying algebraic skills to matrices and systems of equations

Find the solution to a system of equations $Ax = b$, where A is a 3×3 matrix and where the solution is unique	☐
Understand the term 'augmented matrix'	☐
Show that a system of equations has no solutions (inconsistency)	☐
Show that a system of equations has an infinite number of solutions (redundancy)	☐
Compare the solutions of related systems of two equations in two unknowns, and recognise ill-conditioning	☐
Perform matrix operations (at most, order 3): addition, subtraction, multiplication by a scalar, multiplication of matrices	☐
Know and apply the properties of matrix addition and multiplication: 1 $A + B = B + A$ 2 $AB \neq BA$ (in general) 3 $(A + B) + C = A + (B + C)$ (associativity) 4 $(AB)C = A(BC)$ (associativity) 5 $A(B + C) = AB + AC$ (addition is distributive over multiplication)	☐
Know and apply key properties of the transpose, identity matrix and the inverse: 1 $(a_{ij})'_{m \times n} = (a_{ji})_{n \times m}$ (i.e. rows and columns interchange) 2 $(A')' = A$ 3 $(A + B)' = A' + B'$ 4 $(AB)' = B'A'$ 5 A matrix A is orthogonal if $A'A = I$ 6 The $n \times n$ identity matrix I_n: for any square matrix A, $AI_n = I_nA = A$ 7 $B = A^{-1}$ if $AB = BA = I$ 8 $(AB)^{-1} = B^{-1}A^{-1}$	☐
Find the determinant of a 2×2 matrix and a 3×3 matrix	☐
Determine whether a matrix is singular	☐
Know and apply det $(AB) =$ det A det B	☐

contd

Know and use the inverse of a 2 × 2 matrix	☐
Find the inverse of a 3 × 3 matrix, for example by matrix algebra	☐
Use 2 × 2 matrices to carry out geometric transformations in the x, y plane; the transformations should include rotations, reflections and dilatations	☐
Apply combinations of transformations	☐

1.2 Applying algebraic and geometric skills to vectors

Use a vector product method in three dimensions to find the vector product	☐
Evaluate the scalar triple product $\mathbf{a} \cdot (\mathbf{b} \times \mathbf{c})$	☐
Find the equation of a line in parametric, symmetric or vector form, given suitable defining information	☐
Find the angle between two lines in three dimensions	☐
Determine whether or not two lines intersect and, where possible, find the point of intersection	☐
Find the equation of a plane in vector form, parametric form or Cartesian form, given suitable defining information	☐
Find the point of intersection of a plane with a line which is not parallel to the plane	☐
Determine the intersection of two or three planes	☐
Find the angle between a line and a plane or between two planes	☐

1.3 Applying geometric skills to complex numbers

Plot complex numbers in the complex plane (an Argand diagram)	☐
Know the definition of modulus and argument of a complex number	☐
Convert a given complex number from Cartesian to polar form and vice versa	☐
Use de Moivre's theorem with integer and fractional indices	☐
Apply de Moivre's theorem to multiple angle trigonometric formulae	☐
Apply de Moivre's theorem to find the nth roots of a complex number	☐
Interpret geometrically certain equations or inequalities in the complex plane, i.e. find the loci defined by (in)equalities	☐

1.4 Applying algebraic skills to number theory

Use Euclid's algorithm to find the greatest common divisor of two positive integers, i.e. use the division algorithm repeatedly	☐
Express the greatest common divisor (of two positive integers) as a linear combination of the two	☐
Express integers in bases other than ten	☐
Know and use the fundamental theorem of arithmetic	☐

1.5 Applying algebraic and geometric skills to methods of proof

Disprove a conjecture by providing a counter example	☐
Know and be able to use the symbols ∃ and ∀	☐
Write down the negation of a statement	☐
Prove a statement by contradiction	☐
Use further proof by contradiction	☐
Use proof by contrapositive	☐
Use direct proof in straightforward examples	☐

ANSWERS

1. ALGEBRA

Partial fractions – pp. 10-11

1 $\dfrac{x-2}{3x^2+10x+3} = \dfrac{x-2}{(3x+1)(x+3)} = \dfrac{A}{3x+1} + \dfrac{B}{x+3}$

$x - 2 = A(x+3) + B(3x+1);$

$x = -3 \Rightarrow -5 = -8B,\ B = \dfrac{5}{8}$

$x = -\dfrac{1}{3} \Rightarrow -2\dfrac{1}{3} = 2\dfrac{2}{3}A,\ A = -\dfrac{7}{8}$ or

Or, equating coefficients of x^0: $= 3A + B = -2$

$x: A + 3B = 1$

and solve simultaneously to get A and B.

$A = -\dfrac{7}{8} \quad B = \dfrac{5}{8}$

$\therefore \dfrac{x-2}{3x^2+10x+3} = \dfrac{5}{8(x+3)} - \dfrac{7}{8(3x+1)}$

2 $\dfrac{1}{x^3+4x} = \dfrac{1}{x(x^2+4)} = \dfrac{A}{x} + \dfrac{Bx+C}{x^2+4}$

$1 = A(x^2+4) + Bx^2 + Cx;\ x = 0 \Rightarrow 1 = 4A,\ A = \dfrac{1}{4}$

Equating coefficients of x^2 and x: $0 = A + B, 0 = C$.

$A = \dfrac{1}{4},\ B = -\dfrac{1}{4},\ C = 0.$

$\therefore \dfrac{1}{x^3+4x} = \dfrac{1}{4x} - \dfrac{x}{4(x^2+4)}$

3   Dividing out, $\dfrac{x^3+x^2+2}{(x+1)^2} = x - 1 + \dfrac{x+3}{(x+1)^2}$

Let $\dfrac{x+3}{(x+1)^2} = \dfrac{A}{x+1} + \dfrac{B}{(x+1)^2}$

$x + 3 = A(x+1) + B;\ x = -1$ gives $B = 2$; coefficient of x gives $A = 1$

Hence, $\dfrac{x^3+x^2+2}{(x+1)^2} = x - 1 + \dfrac{1}{x+1} + \dfrac{2}{(x+1)^2}$

4 $\dfrac{3x^2-13x+50}{(x-2)(x-3)} = \dfrac{3x^2-13x+50}{x^2-5x+6} = 3 + \dfrac{2x+32}{(x-2)(x-3)}$ by division

$\dfrac{2x+32}{(x-2)(x-3)} = \dfrac{A}{x-2} + \dfrac{B}{x-3}$

$2x + 32 = A(x-3) + B(x-2)$

$x = 2: 36 = -A \Rightarrow A = -36$

$x = 3: 38 = B$

$\dfrac{3x^2-13x+50}{(x-2)(x-3)} = 3 - \dfrac{36}{x-2} + \dfrac{38}{x-3}$

5 $\dfrac{2x^3+3x^2-14x+k}{x^2+4x+3} = 2x - 5 + \dfrac{k+15}{x^2+4x+3}$ by division

As $x \to \infty$, $\dfrac{k+15}{x^2+4x+3} \to 0$,

so $2x - 5 + \dfrac{k+15}{x^2+4x+3} \to 2x - 5$

So $f(x) = \dfrac{2x^3+3x^2-14x+k}{x^2+4x+3}$ approaches the line

$y = 2x - 5$ from above.

6

$$x^3 + 0x + x \overline{)\, x^3 + 0x^2 + 0x + 2}$$

$$\underline{\quad x^3 \qquad\quad +x \quad}$$

$$\qquad\qquad -x + 2$$

$\therefore \dfrac{x^3+2}{x(x^2+1)} = 1 + \dfrac{2-x}{x(x^2+1)}$

$\dfrac{2-x}{x(x^2+1)} \equiv \dfrac{A}{x} + \dfrac{Bx+C}{x^2+1}$

$\therefore 2 - x \equiv A(x^2+1) + x(Bx+C)$

Comparing coefficients:

$x^2: \ 0 = A + B$

$x: \ -1 = C$

$c: \ 2 = A \qquad\qquad \therefore B = -2$

$\therefore \dfrac{x^3+2}{x(x^2+1)} = 1 + \dfrac{2}{x} - \dfrac{2x+1}{x^2+1}$

Factorials and binomial coefficients – pp. 12-13

1 $\dbinom{n+1}{4} - \dbinom{n}{4} = \dfrac{(n+1)!}{4!\,(n-3)!} - \dfrac{n!}{4!\,(n-4)!}$

$= \dfrac{n!}{4!\,(n-4)!}\left[\dfrac{n+1}{n-3} - 1\right]$

$= \dfrac{n!}{4!\,(n-4)!}\left[\dfrac{n+1-(n-3)}{n-3}\right]$

$= \dfrac{n!}{4!\,(n-4)!}\left[\dfrac{4}{n-3}\right]$

$= \dfrac{4 \times n!}{4!\,(n-3)!\,(n-4)!}$

$= \dfrac{n!}{3!\,(n-3)!}$

$= \dbinom{n}{3}$

2 $8\dbinom{n}{2} = 3\dbinom{n}{3}$

$8 \times \dfrac{n!}{(n-2)!\,2!} = 3 \times \dfrac{n!}{(n-3)!\,3!}$

$\dfrac{8 \times (n)(n-1)\cancel{(n-2)!}}{\cancel{(n-2)!}\,2!} = \dfrac{3 \times (n)(n-1)(n-2)\cancel{(n-3)!}}{\cancel{(n-3)!}\,3 \times 2!}$

$\dfrac{8n(n-1)}{2} = \dfrac{n(n-1)(n-2)}{2}$

$8 = n - 2 \Rightarrow n = 10$

3 $\dbinom{n+1}{p+1} - \dbinom{n}{p+1}$

$= \dfrac{(n+1)!}{[(n+1)-(p+1)]!(p+1)!} - \dfrac{n!}{[n-(p+1)]!(p+1)!}$

$= \dfrac{(n+1)\,n!}{(n-p)!(p+1)!} - \dfrac{n!}{(n-p-1)!(p+1)!}$

$= \dfrac{n!}{(p+1)!}\left[\dfrac{(n+1)}{(n-p)!} - \dfrac{1}{(n-p-1)!}\right]$

$= \dfrac{n!}{(p+1)\,p!}\left[\dfrac{(n+1)}{(n-p)!} - \dfrac{(n-p)}{(n-p)!}\right]$

$= \dfrac{n!}{(p+1)\,p!}\left(\dfrac{p+1}{(n-p)!}\right)$

$= \dfrac{n!}{(n-p)\,p!}$

$= \dbinom{n}{p}$ as required.

4 $\dfrac{n(n-1)}{2} + (n+1) = \dfrac{2(n-1)(n-2)}{2} - 1$

$\therefore n(n-1) + 2(n+1) = 2(n-1)(n-2) - 2$

$\therefore n^2 - n + 2n + 2 = 2n^2 - 6n + 4 - 2$

$\therefore 0 = n^2 - 7n$

$\therefore n(n-7) = 0$

contd

$\therefore n = 0$ or $n - 7 = 0$

but $n \geqslant 3 \therefore n = 7$

The binomial theorem – pp. 14-15

1 $(a^3 + 2)^4 = (a^3)^4 + \binom{4}{1}(a^3)^3 \, 2 + \binom{4}{2}(a^3)^2 \, 2^2 + \binom{4}{3}(a^3)2^3 + 2^4$

$= a^{12} + 8a^9 + 24a^6 + 32a^3 + 16.$

2 $\left(2x^3 + \frac{1}{x}\right)^{15} = \sum_{r=0}^{15} \binom{15}{r}\left(\frac{1}{x}\right)^{15-r}(2x^3)^r$

General term $\binom{15}{r}\left(\frac{1}{x}\right)^{15-r}(2x^3)^r$

$= \binom{15}{r} x^{r-15} \, 2^r \, x^{3r}$

$= \binom{15}{r} 2^r \, x^{4r-15}$

For term in x^5: $4r - 15 = 5$

$\therefore r = 5$

$\binom{15}{5} 2^5 \, x^5$

$= \frac{15 \times 14 \times 13 \times 12 \times 11}{5 \times 4 \times 3 \times 2 \times 1} \times 32x^5$

$= 96\,096 \, x^5$

3 General term: $\binom{7}{r}(18x^2)^{7-r}(-1)^r \left(\frac{1}{12}\right)^r \left(\frac{1}{x}\right)^r$

$= \binom{7}{r} 18^{7-r} x^{2(7-r)} (-1)^r \frac{1}{12^r} x^{-r}$

$= \binom{7}{r} 2^{7-r} 3^{2(7-r)} x^{14-2r} (-1)^r \frac{1}{3^r \, 2^{2r}} x^{-r}$

$= \binom{7}{r} 2^{7-r} 3^{14-2r} x^{14-3r} (-1)^r 3^{-r} 2^{-2r}$

$= \binom{7}{r} 2^{7-3r} 3^{14-3r} (-1)^r x^{14-3r}$

For x^5: $14 - 3r = 5 \therefore r = 3$

$\therefore$ term in $x^5 = \binom{7}{3} 2^{-2} 3^5 (-1)^3 x^5$

$= \frac{7 \times 6 \times 5}{3 \times 2 \times 1} \cdot \frac{1}{4} \cdot 243 \times (-1) \, x^5$

$= -\frac{8505}{4} x^5$

4 General term is $\binom{6}{r}(3x)^r \left(-\frac{2}{x^2}\right)^{6-r} = \binom{6}{r} 3^r (-2)^{6-r} x^{3r-12}$.

Need x^0 so $3r - 12 = 0$ and $r = 4$. Constant term

$\binom{6}{4} 3^4 (-2)^2 = \frac{6 \times 5}{2} \times 81 \times 4 = 4860.$

5 Coefficient of x^r in $(1+ x)^{n+1}$ is $\binom{n+1}{r}$

RHS is $(1+ ... + \binom{n}{r-1}x^{r-1} + \binom{n}{r}x^r + ...)(1+ x)$

so coefficient of x^r is $\binom{n}{r} + \binom{n}{r-1}$.

Only use this method when asked to.

6 From $\binom{n+1}{r} = \binom{n}{r} + \binom{n}{r-1} = \binom{n}{r-1} + \binom{n}{r}$

we have $\binom{n+1}{r+1} = \binom{n}{r} + \binom{n}{r+1}$

$(r \to r+1)$,

and $\binom{n+1}{r+2} = \binom{n}{r+1} + \binom{n}{r+2}$ $(r \to r+2)$.

Adding these two results gives

$\binom{n}{r} + 2\binom{n}{r+1} + \binom{n}{r+2} = \binom{n+1}{r+1} + \binom{n+1}{r+2} = \binom{n+2}{r+2}$.

7 Set $x = 1$ in $(1+ x)^n = 1+ \binom{n}{1}x + ... + \binom{n}{n-1}x^{n-1} + x^n$.

$2^n = 1 + \binom{n}{1}(1) + \binom{n}{2}(1)^2 + ... + \binom{n}{n-1}(1)^{n-1} + 1^n$

$= \binom{n}{1} + \binom{n}{2} + ... + \binom{n}{n-1} + 1 + 1$

So $\binom{n}{1} + \binom{n}{2} + ... + \binom{n}{n-1} = 2^n - 2$

as required.

---

2. DIFFERENTIATION

Standard differentials and rules for differentiation – pp. 16-17

1 $f'(x) = 2 \sin x \cos x e^{-\tan x} + \sin^2 x(-\sec^2 x)e^{-\tan x}$

$= (\sin 2x - \tan^2 x)e^{-\tan x}$

$f'\left(\frac{\pi}{4}\right) = \left(\sin \frac{\pi}{2} - \tan^2\left(\frac{\pi}{4}\right)\right)e^{-\tan \frac{\pi}{4}} = (1- 1)e^{-1} = 0.$

2 $f'(x) = 2x \tan 3x + x^2 3 \sec^2 3x$

$= x(2 \tan 3x + 3x \sec^2 3x)$

$= x \sec x(2 \sin 3x + 3x \sec 3x)$

3 $\frac{dy}{dx} = \frac{(1 + 2x)e^x - 2e^x}{(1 + 2x)^2}$

$= \frac{(2x - 1)e^x}{(1 + 2x)^2}$

$\frac{dy}{dx} = 0$ when $(2x - 1)e^x = 0$, $e^x \neq 0$, $\therefore x = \frac{1}{2}$.

4 $S(t) = t^4 - 2t^3 + \frac{3}{2}t^2$

$S'(t) = 4t^3 - 6t^2 + 3t$

$S''(t) = 12t^2 - 12t + 3$

Constant velocity $\therefore 12t^2 - 12t + 3 = 0$

$\therefore 4t^2 - 4t + 1 = 0$

$\therefore (2t - 1)^2 = 0$

$\therefore t = \frac{1}{2}$

$\therefore$ First constant velocity occurs after $\frac{1}{2}$ second.

contd

ANSWERS (Contd.)

Further differentiation 1 - pp. 18-19

1 $\tan y = x$

$\therefore \sec^2 y = \frac{dx}{dy}$

$\therefore \frac{dy}{dx} = \cos^2 y$:

$\therefore \frac{dy}{dx} = \cos^2(\tan^{-1} x)$

2 $\cos y = e^x - 2x$

$\therefore -\sin y \frac{dy}{dx} = e^x - 2$

$\therefore \frac{dy}{dx} = -\frac{e^x - 2}{\sin y}$

$\therefore$ for stationary points $e^x - 2 = 0$

$\therefore e^x = 2$

$\therefore x = \ln 2$

3 $f'(x) = \frac{1}{\sqrt{1 - (4x)^2}} \cdot 4$

$= \frac{4}{\sqrt{1 - 16x^2}}$

Further differentiation 2 - pp. 20-21

1 (a) $x\frac{dy}{dx} + y + 2y\frac{dy}{dx} = 0$

$\therefore \frac{dy}{dx}(x + 2y) = -y$

$\therefore \frac{dy}{dx} = -\frac{y}{x + 2y}$

(b) $\frac{d^2y}{dx^2} = \frac{-(x + 2y)\frac{dy}{dx} + y\left(1 + 2\frac{dy}{dx}\right)}{(x + 2y)^2}$

$= \frac{y - x\left(\frac{-y}{x + 2y}\right)}{(x + 2y)^2}$

$= \frac{y(x + 2y) + xy}{(x + 2y)^2}$

$= \frac{2y(x + y)}{(x + 2y)^3}$

2 Differentiating wrt x: $2y\frac{dy}{dx} + 3y + 3x\frac{dy}{dx} = 2x$

$\frac{dy}{dx} = \frac{2x - 3y}{2y + 3x}$

At $P(-1, 1)$: $\frac{dy}{dx} = \frac{-2 - 3}{2 - 3} = 5$,

so the gradient of the tangent is 5.

or

If the gradient of the tangent to the curve at P is m,

$2(1)m + 3(1) + 3(-1)(1)m = -2$,

i.e. $-m = -5 \Rightarrow m = 5$.

3 $e^{xy} = y \sin x$

$\therefore \left(x\frac{dy}{dx} + y\right)e^{xy} = y \cos x + \frac{dy}{dx}\sin x$

$\therefore \frac{dy}{dx}(xe^{xy} - \sin x) = y \cos x - ye^{xy}$

$\therefore \frac{dy}{dx} = \frac{y \cos x - ye^{xy}}{xe^{xy} - \sin x}$

4 $\ln y = \ln\left[\frac{(2x - 1)^3}{(x - 2)^2}\right]$

$= 3 \ln(2x - 1) - 2 \ln(x - 2)$

$\therefore \frac{1}{y}\frac{dy}{dx} = \frac{3 \times 2}{2x - 1} - \frac{2}{x - 2}$

$\therefore \frac{dy}{dx} = \left(\frac{6}{2x - 1} - \frac{2}{x - 2}\right)\frac{(2x - 1)^3}{(x - 2)^2}$

$= \frac{6(x - 2) - 2(2x - 1)}{(2x - 1)(x - 2)} \cdot \frac{(2x - 1)^3}{(x - 2)^2}$

$= \frac{6x - 12 - 4x + 2}{(x - 2)} \cdot \frac{(2x - 1)^2}{(x - 2)^2}$

$= \frac{2x - 10}{(x - 2)} \cdot \frac{(2x - 1)^2}{(x - 2)^2}$

$= \frac{2(x - 5)(2x - 1)^2}{(x - 2)^3}$

5 (a) $f'(x) = 3 \sin 3x \exp(-\cos 3x)$

(b) $\ln y = \ln 2^{(x^3 + x)}$

$\ln y = (x^3 + x)\ln 2$

$\frac{1}{y}\frac{dy}{dx} = (3x^2 + 1)\ln 2$

so $\frac{dy}{dx} = y(3x^2 + 1)\ln 2$

$= \ln 2(3x^2 + 1)2^{x^3 + x}$

Further differentiation 3 - pp. 22-23

1 When depth is x, radius $= x \tan\frac{\pi}{3} = x\sqrt{3}$

so $V = \frac{\pi}{3}\left(x\sqrt{3}\right)^2 x = \pi x^3$

Hence $\frac{dV}{dt} = 3\pi x^2\frac{dx}{dt}$ and we are given $\frac{dV}{dt} = 2$

so $3\pi x^2\frac{dx}{dt} = 2$

When $x = 3$, $3\pi 3^2\frac{dx}{dt} = 2 \Rightarrow \frac{dx}{dt} = \frac{2}{27\pi}$,

so depth is increasing at $\frac{2}{27\pi}$ cm per second.

2 $\frac{dx}{d\theta} = \sec^2\theta$, $\frac{dy}{d\theta} = \cos\theta$

$\frac{dy}{dx} = \frac{dy}{d\theta}\Big/\frac{dx}{d\theta}$

$\Rightarrow \frac{dy}{dx} = \frac{\cos\theta}{\sec^2\theta} = \cos^3\theta$

3. FUNCTIONS

Properties of functions - pp. 24-25

1   $f(-x) = (-x)^3 \tan(-x) = -x^3(-\tan x) = x^3 \tan x$

So $f(x) = f(-x)$ for all x, and so f is even.

2   (a)

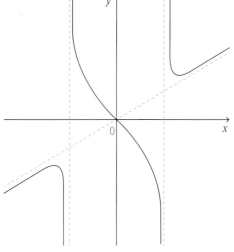

(b) $x = -2$

(c)

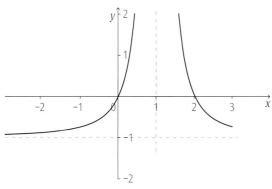

3

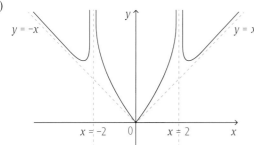

Asymptotes are $y = -1$ and $x = 1$.

4   Neither. Although the graph has half-turn symmetry about A (2, 2), it is not about the origin and therefore is not **odd**. It has no line symmetry and so is not **even**. Therefore, it is **neither** odd nor even.

Critical points 1 - pp. 26-27

1

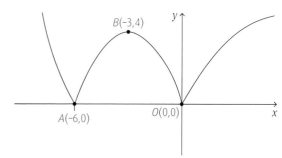

The critical points are A (−6, 0), B (−3, 4) and the origin, O (0, 0).

2

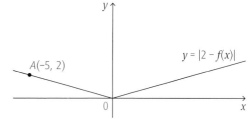

Critical point at the origin, (0, 0).

Critical points 2 - pp. 28-29

1   (a) $f(x) = \frac{x^2 + 3x}{x + 1} = x + 2 - \frac{2}{x + 1}$ using long division,

so there are asymptotes.

Oblique asymptote: $y = x + 2$

and vertical asymptote: $x = -1$.

(b) $f'(x) = 1 + \frac{2}{(x + 1)^2} > 0$ for all $x \neq -1$,

so the graph of f is always increasing.

(c)

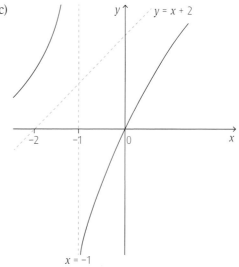

ANSWERS (Contd.)

2 (a) $A: y = 0$ $\therefore \dfrac{kx + 5}{kx - 2} = 0$ $\therefore kx + 5 = 0$ $\therefore x = -\dfrac{5}{k}$

$\therefore A\left(-\dfrac{5}{k}, 0\right)$

$B: x = 0$ $\therefore y = -\dfrac{5}{2}$ $\therefore B\left(0, -\dfrac{5}{2}\right)$

(b) For vertical asymptote: $kx - 2 = 0$ $\therefore x = \dfrac{2}{k}$

For horizontal:

$$\begin{array}{r} 1 \\ kx - 2 \overline{\smash{)}\, kx + 5} \\ \underline{kx - 2} \\ 7 \end{array}$$

$\therefore f(x) = 1 + \dfrac{7}{kx - 2}$

$\therefore y = 1$

(c) $f(x) = 1 + \dfrac{7}{kx - 2} = 1 + 7(kx - 2)^{-1}$

$\therefore f'(x) = -7(kx - 2)^{-2}k$

$= -\dfrac{7k}{(kx - 2)^{-2}}$

And since $k > 0$ and $x \neq \dfrac{2}{k}$,

$\dfrac{7k}{(kx - 2)^{-2}} > 0$

$\therefore -\dfrac{7k}{(kx - 2)^2} < 0$

$\therefore$ Gradient is always negative and so there are no stationary values.

3 (a) $y = \dfrac{x - 3}{x + 2} = 1 - \dfrac{5}{x + 2}$

Vertical asymptote is $x = -2$.

Horizontal asymptote is $y = 1$.

(b) $\dfrac{dy}{dx} = \dfrac{5}{(x + 2)^2} \neq 0$

so no stationary values.

(c) $\dfrac{d^2y}{dx^2} = \dfrac{-10}{(x + 2)^3} \neq 0$

so no points of inflexion.

(d)

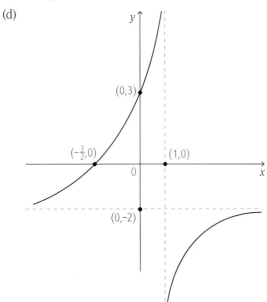

The asymptotes are $x = 1$ and $y = -2$. The domain must exclude $x = 1$.

NB. You are not required to obtain a formula for f^{-1}.

Related graphs – pp. 30–31

1 (a)

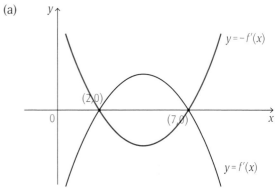

(b)

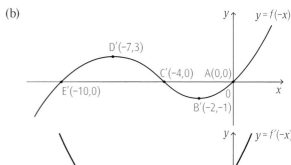

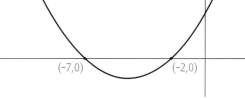

Graphs of trigonometric functions – pp. 32–33

1 Range in two parts: $y \geqslant 1$; $y \leqslant -1$.

2 Vertical asymptotes at $x = \pm\,\pi$, $x = 0$.

3 Rotational symmetry about $(0, 0)$.

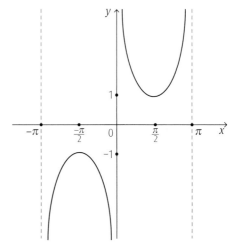

4. INTEGRATION

Basic integration and integration by substitution - pp. 34-35

1 $x = (u-1)^2 \Rightarrow dx = 2(u-1)du$

$\int \frac{1}{(1+\sqrt{x})^3}dx = \int \frac{2(u-1)}{u^3}du$

$\qquad = 2\int (u^{-2} - u^{-3})du$

$\qquad = 2\left(\frac{-1}{u} + \frac{1}{2u^2}\right) + c = \frac{1}{(1+\sqrt{x})^2} - \frac{2}{1+\sqrt{x}} + c$

2 (a) $1 + x^2 = u \Rightarrow 2x\,dx = du,\ x\,dx = \frac{du}{2}$

$\quad x = 0 \Rightarrow u = 1;\ x = 1 \Rightarrow u = 2$

$\quad \int_0^1 \frac{x^3}{(1+x^2)^4}dx = \int_1^2 \frac{(u-1)}{2u^4}du$

$\quad = \frac{1}{2}\int_1^2 (u^{-3} - u^{-4})\,du$

$\quad = \frac{1}{2}\left[-\frac{1}{2}u^{-2} + \frac{1}{3}u^{-3}\right]_1^2 = \frac{1}{24}$

(b) Hence the volume of revolution $= \frac{\pi}{24}$

3 $x = 2\sin\theta \Rightarrow dx = 2\cos\theta\,d\theta$

$x = 0 \Rightarrow \theta = 0;\ x = \sqrt{2} \Rightarrow \sin\theta = \frac{1}{\sqrt{2}} \Rightarrow \theta = \frac{\pi}{4}$

$\int_0^{\sqrt{2}} \frac{x^2}{\sqrt{4-x^2}}dx = \int_0^{\pi/4} \frac{4\sin^2\theta}{\sqrt{4-4\sin^2\theta}}(2\cos\theta)d\theta$

$= \int_0^{\pi/4} \frac{4\sin^2\theta}{2\cos\theta}(2\cos\theta)d\theta$

$= 2\int_0^{\pi/4} (2\sin^2\theta)d\theta$

$= 2\int_0^{\pi/4} (1-\cos 2\theta)d\theta$

$= 2\left[\theta - \frac{1}{2}\sin 2\theta\right]_0^{\pi/4}$

$= 2\left\{\left[\frac{\pi}{4} - \frac{1}{2}\right] - 0\right\}$

$= \frac{\pi}{2} - 1$

Further integration - pp. 36-37

1 (a) $\frac{1}{x^2 + 2x - 8} = \frac{1}{(x-2)(x+4)} = \frac{A}{x-2} + \frac{B}{x+4}$

$\qquad 1 = A(x+4) + B(x-2);$

$\quad x = 2 \Rightarrow A = \frac{1}{6};$

$\quad x = -4 \Rightarrow B = -\frac{1}{6}$

$\quad$ So $\frac{1}{x^2} + 2x - 8 = \frac{1}{6}(x-2) - \frac{1}{6}(x+4)$

(b) $\int_0^1 \frac{1}{x^2+2x-8}dx = \frac{1}{6}\int_0^1 \left(\frac{1}{x-2} - \frac{1}{x+4}\right)dx$

$\qquad = \frac{1}{6}\left[\ln|x-2| - \ln|x+4|\right]_0^1$

$\qquad = \frac{1}{6}\left[\ln 1 - \ln 5 - \ln 2 + \ln 4\right]$

$\qquad = \frac{1}{6}(\ln 2 - \ln 5)$

$\qquad = \frac{1}{6}\ln\frac{2}{5}$

2 (a) $\frac{1}{x^3 + x} = \frac{1}{x} - \frac{x}{x^2+1}$

(b) $I(k) = \int_1^k \frac{1}{x^3+x}dx = \int_1^k \left(\frac{1}{x} - \frac{x}{x^2+1}\right)dx$

$\qquad = \int_1^k \frac{1}{x}dx - \int_1^k \frac{x}{x^2+1}dx$

$\qquad = \left[\ln x\right]_1^k - \frac{1}{2}\left[\ln(x^2+1)\right]_1^k$

$\qquad = \left[\ln k - 0\right] - \frac{1}{2}\left[\ln(k^2+1) - \ln 2\right]$

$\qquad = \ln k - \ln\sqrt{k^2+1} + \ln\sqrt{2}$

$\qquad = \ln\frac{k\sqrt{2}}{\sqrt{k^2+1}}$

3 $V = \pi\int_1^2 y^2\,dx = \pi\int_1^2 (x^2-1)^2\,dx$

$\qquad = \pi\int_1^2 (x^4 - 2x^2 + 1)\,dx$

$\qquad = \pi\left[\frac{1}{5}x^5 - \frac{2}{3}x^3 + x\right]_1^2$

$\qquad = \pi\left(\frac{32}{5} - \frac{2}{3}\cdot 8 + 2\right) - \left(\frac{1}{5} - \frac{2}{3} + 1\right)$

$\qquad = \pi\left[\frac{32}{5} - \frac{16}{3} + 2 - \frac{1}{5} + \frac{2}{3} - 1\right]$

$\qquad = \pi\left[\frac{31}{5} - \frac{14}{3} + 1\right] = 2\frac{8}{15}\pi\,\text{units}^3$

4 $V = \pi\int_0^3 x^2\,dy$

$\qquad = \pi\int_0^3 (y+1)^{\frac{1}{2}}\,dy$

$\qquad = \pi\left[\frac{2}{3}(y+1)^{\frac{3}{2}}\right]_0^3$

$\qquad = \pi\left[\frac{16}{3} - \frac{2}{3}\right]$

$\qquad = \frac{14}{3}\pi\,\text{units}^3$

$\qquad y = x^4 - 1$

$\qquad \therefore y + 1 = x^4$

$\qquad \therefore \sqrt{y+1} = x^2$

Integration by parts - pp. 38-39

1 $\int 10x^2\cos 5x\,dx = 10x^2\frac{\sin 5x}{5} - \int 20x\frac{\sin 5x}{5}dx$

$\qquad = 2x^2\sin 5x - 4\int x\sin 5x\,dx$

$\int x\sin 5x\,dx = x\frac{\cos 5x}{-5} + \int \frac{\cos 5x}{5}dx$

$\qquad = -\frac{1}{5}x\cos 5x + \frac{1}{25}\sin 5x$

So $\int 10x^2\cos 5x\,dx$

$\qquad = 2x^2\sin 5x + \frac{4}{5}x\cos 5x - \frac{4}{25}\sin 5x + c$

2 $\int_0^1 x\tan^{-1}x^2\,dx = \left[\tan^{-1}x^2\int x\,dx\right]_0^1 - \int_0^1 \frac{2x}{1+x^4}\frac{x^2}{2}dx$

$\qquad = \left[\frac{1}{2}x^2\tan^{-1}x^2\right]_0^1 - \int_0^1 \frac{x^3}{1+x^4}dx$

$\qquad = \left[\frac{1}{2}x^2\tan^{-1}x^2\right]_0^1 - \left[\frac{1}{4}\ln(1+x^4)\right]_0^1$

$\qquad = \frac{1}{2}\tan^{-1}1 - 0 - \left[\frac{1}{4}\ln 2 - \frac{1}{4}\ln 1\right]$

$\qquad = \frac{\pi}{8} - \frac{1}{4}\ln 2$

contd

ANSWERS (Contd.)

3 $\int \frac{(x+3)^2}{(x+1)^3}\,dx = -\frac{(x+3)^2}{2(x+1)^2} + \int \frac{x+3}{(x+1)^2}\,dx$

$\int \frac{x+3}{(x+1)^2}\,dx = -\frac{x+3}{x+1} + \int \frac{1}{x+1}\,dx$

so $\int \frac{(x+3)^2}{(x+1)^3}\,dx = -\frac{(x+3)^2}{2(x+1)^2} - \frac{x+3}{x+1} + \ln|x+1| + c$.

4 $I = \int_0^{\pi/2} e^{2x}\cos x\,dx = \left[e^{2x}\sin x\right]_0^{\pi/2} - \int_0^{\pi/2} 2e^{2x}\sin x\,dx$

$= [e^\pi - 0] - 2\int_0^{\pi/2} e^{2x}\sin x\,dx$

$= e^\pi - 2J$

$J = \int_0^{\pi/2} e^{2x}\sin x\,dx = \left[-e^{2x}\cos x\right]_0^{\pi/2} + \int_0^{\pi/2} 2e^{2x}\cos x\,dx$

$= [0 + 1] + 2\int e^{2x}\cos x\,dx$

$= 1 + 2I$

so $I = e^\pi - 2J$ and $J = 1 + 2I$.

Hence $I = e^\pi - 2(1 + 2I)$

giving

$I = \frac{1}{5}(e^\pi - 2)$.

5. EQUATIONS

Systems of linear equations and Gaussian Elimination – pp. 40–41

1

R_1	1	3	5	14
R_2	2	−1	−3	3
R_3	4	5	−1	7
R_1	1	3	5	14
R_4	0	7	13	25	$2R_1 - R_2$
R_5	0	7	21	49	$4R_1 - R_3$
R_1	1	3	5	14
R_4	0	7	13	25
R_6	0	0	8	24	$R_5 - R_4$

$8z = 24 \Rightarrow z = 3$

$7y + 13(3) = 25 \Rightarrow y = -2$

$x + 3(-2) + 5(3) = 14 \Rightarrow x = 5$

2

R_1	1	1	−1	6
R_2	2	−3	2	2
R_3	−5	2	−4	1
R_1	1	1	−1	6
R_4	0	−5	4	−10	$R_2 - 2R_1$
R_5	0	7	−9	31	$R_3 + 5R_1$
R_1	1	1	−1	6
R_4	0	−5	4	−10	$5R_5 + 7R_4$
R_6	0	0	−17	85

$-17z = 85 \Rightarrow z = -5$

$-5y - 20 = -10 \Rightarrow y = -2$

$x - 2 + 5 = 6 \Rightarrow x = 3$

Special cases – pp. 42–43

1

R_1	1	3	2	4
R_2	2	4	−5	−8
R_3	1	−3	−25	−44
R_1	1	3	2	4
R_4	0	2	9	16	$2R_1 - R_2$
R_5	0	6	27	48	$R_1 - R_3$
R_1	1	3	2	4
R_4	0	2	9	16
R_6	0	0	0	0	$R_5 - 3R_4$

The third row is redundant – the third equation is redundant. The system has an infinite number of solutions.

Let $z = t$

$2y + 9t = 16$

$y = \frac{16 - 9t}{2}$

$x + 3\left(\frac{16 - 9t}{2}\right) + 2t = 4$

$x = \frac{23t - 40}{2}$

2

1	1	2	3	
2	3	4	5	$R_2 \rightarrow R_2 - 2R_1$
3	2	λ	11	$R_3 \rightarrow R_3 - 3R_1$
1	1	2	3	
0	1	0	−1	R_2
0	−1	$\lambda - 6$	2	R_3

$\therefore y = -1$ $R_3 \rightarrow 1 + (\lambda - 6)z = 2$

$\therefore z = \frac{1}{\lambda - 6}$

$x + y + 2z = 3 \rightarrow x - 1 + \frac{2}{\lambda - 6} = 3$

$\therefore x = 4 - \frac{2}{\lambda - 6}$

contd

When $\lambda = 6$, the system is inconsistent so has no solutions [R_2: $y = -1$ R_3: $y = -2$]

3 (a)
$$\begin{array}{ccc|c}
1 & -1 & 1 & -7 \\
1 & 1 & 3 & -1 \\
3 & 1 & \lambda & -9
\end{array} \quad \begin{array}{l} R_2 \to R_2 - R_1 \\ R_3 \to R_3 - 3R_1 \end{array}$$

$$\begin{array}{ccc|c}
1 & -1 & 1 & -7 \\
0 & 2 & 2 & 6 \\
0 & 4 & \lambda - 3 & 12
\end{array} \quad \begin{array}{l} R_2 \to \frac{1}{2}R_2 \\ R_3 \to R_3 - 2R_2 \end{array}$$

$$\begin{array}{ccc|c}
1 & -1 & 1 & -7 \\
0 & 1 & 1 & 3 \\
0 & 0 & \lambda - 7 & 0
\end{array}$$

$\lambda \neq 7$: $z = 0$, $y = 3$, $x = -4$

(b) R_3: $0\ 0\ 0\,|\,0$ ∴ redundant, infinite solutions

(c) $z = z$; $y = 3 - z$; $x = -7 + y - z$

$\quad\quad = -7 + (3 - z) - z$

$\therefore x = -4 - 2z$

6. MATRICES

Terms and basic matrix operations - pp. 44–45

1 $\quad AB = \begin{bmatrix} 2 & -3 \\ 1 & 4 \end{bmatrix} \begin{bmatrix} -1 & 5 \\ 0 & -2 \end{bmatrix}$

$= \begin{bmatrix} (2 \times -1) + (-3 \times 0) & (2 \times 5) + (-3 \times -2) \\ (1 \times -1) + (4 \times 0) & (1 \times 5) + (4 \times -2) \end{bmatrix}$

$= \begin{bmatrix} -2 & 16 \\ -1 & -3 \end{bmatrix}$

$BA = \begin{bmatrix} -1 & 5 \\ 0 & -2 \end{bmatrix} \begin{bmatrix} 2 & -3 \\ 1 & 4 \end{bmatrix}$

$= \begin{bmatrix} (-1 \times 2) + (5 \times 1) & (-1 \times -3) + (5 \times 4) \\ (0 \times 2) + (-2 \times 1) & (0 \times -3) + (-2 \times 4) \end{bmatrix}$

$= \begin{bmatrix} 3 & 23 \\ -2 & -8 \end{bmatrix}$

2 $\quad CD = \begin{bmatrix} 2 & 3 & 4 \\ 1 & 0 & -1 \\ 3 & -2 & -1 \end{bmatrix} \begin{bmatrix} 6 & 0 \\ -1 & 2 \\ 4 & -3 \end{bmatrix}$

$= \begin{bmatrix} (2 \times 6 + 3 \times -1 + 4 \times 4) & (2 \times 0 + 3 \times 2 + 4 \times -3) \\ (1 \times 6 + 0 \times -1 + -1 \times 4) & (1 \times 0 + 0 \times 2 + -1 \times -3) \\ (3 \times 6 + -2 \times -1 + -1 \times 4) & (3 \times 0 + -2 \times 2 + -1 \times -3) \end{bmatrix}$

$= \begin{bmatrix} 25 & -6 \\ 2 & 3 \\ 16 & -1 \end{bmatrix}$

Matrix multiplication and the transpose of a matrix - pp. 46–47

1 (a) $B^{T}A = \begin{bmatrix} 1 & b \\ -2b & 1 \end{bmatrix} \begin{bmatrix} 2 & 5 \\ -3 & 7 \end{bmatrix} = \begin{bmatrix} 2 - 3b & 5 + 7b \\ -4b - 3 & -10b + 7 \end{bmatrix}$

(b) $\begin{bmatrix} 2 - 3b & 5 + 7b \\ -4b - 3 & -10b + 7 \end{bmatrix} = \begin{bmatrix} 5 & q \\ p & 17 \end{bmatrix}$

$2 - 3b = 5 \quad\quad \therefore b = -1$

$5 + 7b = q \quad\quad \therefore 5 - 7 = q \quad \therefore q = -2$

$-4b - 3 = p \quad \therefore 4 - 3 = p \quad \therefore p = 1$

2 (a) $3A - 5B = \begin{bmatrix} 6 & 9 & 3 \\ 3p & -3 & 0 \\ 6 & 3 & 3 \end{bmatrix} - \begin{bmatrix} 5 & 5 & 10 \\ 10 & -5 & 0 \\ 0 & 10 & -10 \end{bmatrix}$

$= \begin{bmatrix} 1 & 4 & -7 \\ 3p - 10 & 2 & 0 \\ 6 & -7 & 13 \end{bmatrix}$

(b) $AB = \begin{bmatrix} 2 & 3 & 1 \\ p & -1 & 0 \\ 2 & 1 & 1 \end{bmatrix} \begin{bmatrix} 1 & 1 & 2 \\ 2 & -1 & 0 \\ 0 & 2 & -2 \end{bmatrix} = \begin{bmatrix} 8 & 1 & 2 \\ p - 2 & p + 1 & 2p \\ 4 & 3 & 2 \end{bmatrix}$

$\det(AB) = 8(2p + 2 - 6p) - (2p - 4 - 8p)$
$\quad\quad\quad\quad + 2(3p - 6 - 4p - 4)$

$\quad\quad = 16p + 16 - 48p - 2p + 4 + 8p$
$\quad\quad\quad\quad + 6p - 12 - 8p - 8$

$\quad\quad = -28p$

(c) AB singular if $\det(AB) = 0$

$\quad \therefore 0 = -28p \therefore p = 0$

contd

ANSWERS (Contd.)

Special matrices 1 - pp. 48-49

1. $\det A = \begin{vmatrix} 2 & -3 \\ 1 & 4 \end{vmatrix} = (2 \times 4) - (1 \times -3) = 11$

   $\det B = \begin{vmatrix} -1 & 5 \\ 0 & -2 \end{vmatrix} = (-1 \times -2) - (0 \times 5) = 2$

2. $\det C = 2\begin{vmatrix} 0 & -1 \\ -2 & -1 \end{vmatrix} - 3\begin{vmatrix} 1 & -1 \\ 3 & -1 \end{vmatrix} + 4\begin{vmatrix} 1 & 0 \\ 3 & -2 \end{vmatrix}$

   $= 2((0 \times -1) - (-2 \times -1)) - 3((1 \times -1) - (3 \times -1)) + 4((1 \times -2) - (3 \times 0))$

   $= (2 \times -2) - (3 \times 2) + (4 \times -2)$

   $= -18$

3. (a) $\det \begin{bmatrix} t+4 & 3t \\ 3 & 5 \end{bmatrix} = 5(t+4) - 9t = 20 - 4t$

   $A^{-1} = \frac{1}{20 - 4t}\begin{bmatrix} 5 & -3t \\ -3 & t+4 \end{bmatrix}$

   (b) $20 - 4t = 0 \Rightarrow t = 5$

   (c) $\begin{bmatrix} t+4 & 3 \\ 3t & 5 \end{bmatrix} = \begin{bmatrix} 6 & 3 \\ 6 & 5 \end{bmatrix} \Rightarrow t = 2$

Special matrices 2 - pp. 50-51

1. (a) $A^2 = \begin{bmatrix} 2 & 1 \\ a & 0 \end{bmatrix}\begin{bmatrix} 2 & 1 \\ a & 0 \end{bmatrix} = \begin{bmatrix} 4+a & 2 \\ 2a & a \end{bmatrix}$

   (b) $2A + 3I = \begin{bmatrix} 4 & 2 \\ 2a & 0 \end{bmatrix} + \begin{bmatrix} 3 & 0 \\ 0 & 3 \end{bmatrix} = \begin{bmatrix} 7 & 2 \\ 2a & 3 \end{bmatrix}$

   $= \begin{bmatrix} 4+a & 2 \\ 2a & a \end{bmatrix} \quad \therefore a = 3$

   (c) $A^3 = AA^2 = A(2A + 3I)$

   $= 2A^2 + 3A = 2(2A + 3I) + 3A$

   $= 7A + 6I$

2. $\det\begin{bmatrix} 1 & 2 \\ -x & 3 \end{bmatrix} = 3 + 2x$, so the inverse is $\frac{1}{3+2x}\begin{bmatrix} 3 & -2 \\ x & 1 \end{bmatrix}$

   Matrix is singular when $3 + 2x = 0 \Rightarrow x = -\frac{3}{2}$.

3. (a) $AB = \begin{bmatrix} 6 & -2 & 2 \\ -3 & 1 & -3 \\ 1 & 1 & 1 \end{bmatrix}\begin{bmatrix} 1 & 1 & 1 \\ 0 & 1 & 3 \\ -1 & -2 & 0 \end{bmatrix} = \begin{bmatrix} 4 & 0 & 0 \\ 0 & 4 & 0 \\ 0 & 0 & 4 \end{bmatrix}$

   (b) $AB = 4I \quad \therefore B = 4A^{-1}$

   (c) Since $AB = 4I$, $AB = BA$.

   So $C = BA = AB = 4I = \begin{bmatrix} 4 & 0 & 0 \\ 0 & 4 & 0 \\ 0 & 0 & 4 \end{bmatrix}$

4. Given the matrix $A = \begin{bmatrix} \lambda & 2 \\ \lambda+3 & 4 \end{bmatrix}$

   (a) $\det A = 4\lambda - 2(\lambda + 3) = 2\lambda - 6$,

   so $A^{-1} = \frac{1}{2\lambda - 6}\begin{bmatrix} 4 & -2 \\ -\lambda-3 & \lambda \end{bmatrix}$

   (b) $2\lambda - 6 = 0 \Rightarrow \lambda = 3$

   (c) $A' = \begin{bmatrix} \lambda & \lambda+3 \\ 2 & 4 \end{bmatrix} = \begin{bmatrix} -2 & 1 \\ 2 & 4 \end{bmatrix} \Rightarrow \lambda = -2$.

5. A singular $\Leftrightarrow \det A = 0$

   $\det A = k(-3) - 1(-2) - 1(-4 + 3(k-1))$

   $\therefore -3k + 2 + 4 - 3k + 3 = 0$

   $\therefore -6k + 9 = 0$

   $\therefore k = \frac{3}{2}$

6. (a) $A^3 = 13A - 12I = A^{-1}A^4 = A^{-1}(40A - 39I)$

   $\therefore 13A - 12I = 40I - 39A^{-1}$

   $\therefore 39A^{-1} = 52I - 13A$

   $\therefore A^{-1} = -\frac{13}{39}A + \frac{52}{39}I$

   $\therefore A^{-1} = -\frac{1}{3}A + \frac{4}{3}I$

   (b) $A^2 = A^3A^{-1} = (13A - 12I)\left(-\frac{1}{3}A + \frac{4}{3}I\right)$

   $= -\frac{13}{3}A^2 + \frac{52}{3}A + 4A - 16I$

   $= -\frac{13}{3}A^2 + \frac{64}{3}A - 16I$

   $\therefore \frac{16}{3}A^2 = \frac{64}{3}A - 16I$

   $\therefore A^2 = 4A - 3I$

Linear equations in matrix form and geometric transformations - pp. 52-53

1. (a) $M_1 = \begin{bmatrix} \cos\left(\frac{-\pi}{4}\right) & -\sin\left(\frac{-\pi}{4}\right) \\ \sin\left(\frac{-\pi}{4}\right) & \cos\left(\frac{-\pi}{4}\right) \end{bmatrix} = \begin{bmatrix} \frac{1}{\sqrt{2}} & \frac{1}{\sqrt{2}} \\ -\frac{1}{\sqrt{2}} & \frac{1}{\sqrt{2}} \end{bmatrix}$

   (b) $M_2 = \begin{bmatrix} -1 & 0 \\ 0 & 1 \end{bmatrix}$

   (c) $\begin{bmatrix} -1 & 0 \\ 0 & 1 \end{bmatrix}\begin{bmatrix} \frac{1}{\sqrt{2}} & \frac{1}{\sqrt{2}} \\ -\frac{1}{\sqrt{2}} & \frac{1}{\sqrt{2}} \end{bmatrix} = \begin{bmatrix} -\frac{1}{\sqrt{2}} & -\frac{1}{\sqrt{2}} \\ -\frac{1}{\sqrt{2}} & \frac{1}{\sqrt{2}} \end{bmatrix}$

2. The matrix $\begin{bmatrix} 2 & 0 \\ 0 & 2 \end{bmatrix}$

   gives an enlargement, scale factor 2.

   The matrix $\begin{bmatrix} \frac{1}{2} & \frac{\sqrt{3}}{2} \\ -\frac{\sqrt{3}}{2} & \frac{1}{2} \end{bmatrix}$

   gives a clockwise rotation of 60° about the origin.

   $M = \begin{bmatrix} \frac{1}{2} & \frac{\sqrt{3}}{2} \\ -\frac{\sqrt{3}}{2} & \frac{1}{2} \end{bmatrix}\begin{bmatrix} 2 & 0 \\ 0 & 2 \end{bmatrix} = \begin{bmatrix} 1 & \sqrt{3} \\ -\sqrt{3} & 1 \end{bmatrix}$

7. COMPLEX NUMBERS

Polynomial equations - pp. 54-55

1 Let $z = x + iy$ so equation is $x + iy + x^2 + y^2 = 7 - i$.
Equating real parts: $x + x^2 + y^2 = 7$.

Equating imaginary parts: $y = -1$

Substituting for y into first equation gives

$x^2 + x - 6 = 0$

$(x + 3)(x - 2) = 0$

which gives $x = -3$, $x = 2$.

The solutions are $2 - i$ and $-3 - i$.

2 (a) $\frac{1}{z} = \frac{1}{a + 3i} \times \frac{a - 3i}{a - 3i} = \frac{a - 3i}{a^2 + 9} = \left(\frac{a}{a^2 + 9}\right) - \left(\frac{3}{a^2 + 9}\right)i$

(b) $z\bar{z} = (a + 3i)(a - 3i) = a^2 + 9 = 25$ $\therefore a^2 = 16$

$\therefore a = \pm 4$

3 $3^3 - 9(3)^2 + 28(3) - 30 = 27 - 81 + 84 - 30 = 0$

$\therefore z = 3$ a root

$$
\begin{array}{r}
z^2 - 6z + 10 \\
z - 3 \overline{)\,z^3 - 9z^2 + 28z - 30} \\
\underline{z^3 - 3z^2} \\
-6z^2 + 28z \\
\underline{-6z^2 + 18z} \\
10z - 30 \\
\underline{10z - 30} \\
0
\end{array}
$$

$z^2 - 6z + 10 = 0$

$\therefore (z - 3)^2 = -1$

$\therefore z - 3 = \pm i$

$\therefore z = 3 \pm i$

4 $(2 - 3i)^2 = 4 - 12i + 9i^2 = -5 - 12i$

$(2 - 3i)^3 = (-5 - 12i)(2 - 3i) = -10 + 15i - 24i + 36i^2$
$\qquad = -46 - 9i$

$(2 - 3i)^4 = (-46 - 9i)(2 - 3i) = -92 + 138i - 18i + 27i^2$
$\qquad = -119 + 120i$

$z^4 - 2z^3 + 7z^2 + 18z + 26 = 0$

$\therefore$ LHS $= -119 + 120i - 2(-46 - 9i) + 7(-5 - 12i)$
$\qquad + 18(2 - 3i) + 26$

$\qquad = -119 + 120i + 92 + 18i - 35 - 84i + 36 - 54i + 26$

$\qquad = (-119 + 92 - 35 + 36 + 26) + i(120 + 18 - 84 - 54)$

$\qquad = 0$

$\qquad =$ RHS

$\therefore z = 2 - 3i$ is a solution $\quad \therefore z = 2 + 3i$ is a solution

$(z - 2 + 3i)(z - 2 - 3i) = z^2 - 4z + 13$

$$
\begin{array}{r}
z^2 + 2z + 2 \\
z^2 - 4z + 13 \overline{)\,z^4 - 2z^3 + 7z^2 + 18z + 26} \\
\underline{z^4 - 4z^3 + 13z^2} \\
2z^3 - 6z^2 + 18z \\
\underline{2z^3 - 8z^2 + 26z} \\
2z^2 - 8z + 26 \\
\underline{2z^2 - 8z + 26} \\
0
\end{array}
$$

For other roots, solve

$z^2 + 2z + 2 = 0$

$\therefore (z + 1) - 1 + 2 = 0$

$\therefore (z + 1)^2 = -1 \Rightarrow z + 1 = \pm i$

$\therefore z = -1 \pm i$ $\therefore$ solutions are $z = 2 \pm 3i, -1 \pm i$

5 Since $3 - i$ is a root, so is $3 + i$.

$(z - (3 - i))(z - (3 + i)) = (z - 3 + i)(z - 3 - i)$

$\qquad = (z - 3)^2 + 1$

$\qquad = z^2 - 6z + 10$

Hence $z^2 - 6z + 10$ is a factor of $2z^3 - 11z^2 + 14z + 10$.
By inspection or long division

$2z^3 - 11z^2 + 14z + 10 = (z^2 - 6z + 10)(2z + 1)$,

so the roots are $3 + i, 3 - i, -\frac{1}{2}$.

6 Because $i - 1$ is a root, so is $-i - 1$, and

$(z - i + 1)(z + i + 1) = z^2 + 2z + 2$ is a factor of $f(z)$.

Dividing $z^4 - z^3 - 5z^2 - 8z - 2$ by $z^2 + 2z + 2$ gives

$z^4 - z^3 - 5z^2 - 8z - 2 = (z^2 + 2z + 2)(z^2 - 3z - 1)$, so
other roots are given by $z^2 - 3z - 1 = 0$. This is solved
by the quadratic formula.

The roots are $-i - 1, i - 1, \frac{3}{2} \pm \frac{\sqrt{13}}{2}$

7 (a) $z + |z| = x + iy + \sqrt{x^2 + y^2} = 16 - 8i$

$\therefore x + \sqrt{x^2 + y^2} + iy = 16 - 8i$

Equating imaginary parts: $y = -8$

Equating real parts: $x + \sqrt{x^2 + 64} = 16$

$\therefore \sqrt{x^2 + 64} = 16 - x$

$\therefore x^2 + 64 = 256 - 32x + x^2$

$\therefore 32x = 192$

$\therefore x = 6$

$\therefore z = 6 - 8i$

(b) $z^2 + 3 = 24 - 20i$

$\therefore x^2 + 2ixy - y^2 + 3 = 24 - 20i$

Equating imaginary parts: $2xy = -20$ $\therefore x = -\frac{10}{y}$

contd

ANSWERS (Contd.)

Equating real parts: $x^2 - y^2 + 3 = 24$

$\therefore \frac{100}{y^2} - y^2 - 21 = 0$

$\therefore 100 - y^4 - 21y^2 = 0$

$\therefore y^4 + 21y^2 - 100 = 0$

$\therefore (y^2 + 25)(y^2 - 4) = 0$

$\therefore y = \pm 2, \pm 5i$

$\therefore y = \pm 2 \quad \because y \in \mathbb{R} \quad \therefore x = \mp 5$

$\therefore z = -5 + 2i \quad$ or $\quad 5 - 2i$

Argand diagram and geometric figures – pp.56-57

1 (a) $|z| = 2\sqrt{3}a$; arg $(z) = \frac{\pi}{6}$

   (b) $\therefore z^2 = \left(2\sqrt{3}a\right)^2 \left(\cos \frac{\pi}{6} + i \sin \frac{\pi}{6}\right)^2$

   $= 12a^2 \left(\cos \frac{\pi}{3} + i \sin \frac{\pi}{3}\right)$

   $z^3 = \left(2\sqrt{3}a\right)^3 \left(\cos \frac{\pi}{6} + i \sin \frac{\pi}{6}\right)^3$

   $= 2^3 \left(\sqrt{3}\right)^3 a^3 \left(\cos \frac{\pi}{2} + i \sin \frac{\pi}{2}\right)$

   $= 24\sqrt{3}a^3 \left(\cos \frac{\pi}{2} + i \sin \frac{\pi}{2}\right)$

   (c)

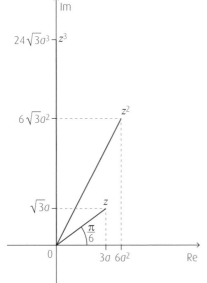

2 (a) $z^2 = \left(\frac{1}{\sqrt{2}} + \frac{1}{\sqrt{2}}i\right)\left(\frac{1}{\sqrt{2}} + \frac{1}{\sqrt{2}}i\right) = \frac{1}{2} + \frac{2}{\sqrt{2}\sqrt{2}}i + \frac{1}{2}i^2$

   $= \frac{1}{2} - \frac{1}{2} + 1i = i$

   $z^3 = i\left(\frac{1}{\sqrt{2}} + \frac{1}{\sqrt{2}}i\right) = \frac{1}{\sqrt{2}}i + \frac{1}{\sqrt{2}}i^2 = -\frac{1}{\sqrt{2}} + \frac{1}{\sqrt{2}}i$

(b)

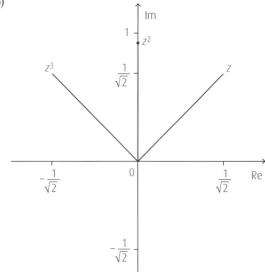

3 (a) $(x + 3)^2 + y^2 = 4^2$

   $\therefore$ Locus is a circle, centre $(-3, 0)$ with radius 4.

   (b) $(x + iy)(x - iy) = x^2 + (y - 2)^2$

   $\therefore x^2 - i^2y = x^2 + y^2 - 4y + 4$

   $\therefore x^2 + y^2 = x^2 + y^2 - 4y + 4$

   $\therefore 0 = -4y + 4$

   $\therefore y = 1$

   $\therefore$ Locus is the straight line $y = 1$.

4 $x^2 + (y - 2)^2 < 3^2$

   Inequality gives a circle centre $(0, 2)$ radius 3 (circumference not included).

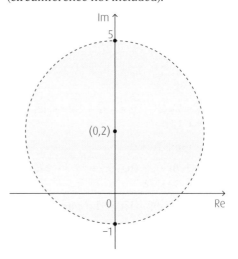

contd

5 $|x + iy - 2i| = |x + iy + 6|$

$\therefore |x + i(y - 2)|^2 = |(x + 6) + iy|^2$

$\therefore x^2 + (y - 2)^2 = (x + 6)^2 + y^2$

$\therefore x^2 + y^2 - 4y + 4 = x^2 + 12x + 36 + y^2$

$\therefore -4y = 12x + 32$

$\therefore y = -3x - 8$

$\therefore$ The locus is given by the straight line $y = -3x - 8$.

Polar form and De Moivre's theorem – pp. 58–59

1 Either $z^3 = zz^2 = (2 + ai)(2 + ai)^2 = (2 + ai)(4 - a^2 + 4ai)$

$\qquad = 8 - 6a^2 + (12a - a^3)i$

or

$z^3 = (2 + ai)^3 = 2^3 + 3(2^2)ai + 3(2)(ai)^2 + (ai)^3$

$\quad = 8 + 12ai - 6a^2 - a^3i = 8 - 6a^2 + (12a - a^3)i$

which is real when $12a - a^3 = 0$.

As $a > 0$, solving gives

$a = \sqrt{12} = 2\sqrt{3}$.

$|2 + i\sqrt{12}| = \sqrt{4 + 12} = 4$,

$\arg\left(2 + i\sqrt{12}\right) = \tan^{-1}\left(\frac{\sqrt{12}}{2}\right) = \tan^{-1}\sqrt{3} = \frac{\pi}{3}$

2 (a) $z^k = (\cos\theta + i\sin\theta)^k$

$\qquad = \cos k\theta + i\sin k\theta$

(b) $z^{-k} = \dfrac{1}{\cos k\theta + i\sin k\theta} = \dfrac{\cos k\theta - i\sin k\theta}{(\cos k\theta + i\sin k\theta)(\cos k\theta - i\sin k\theta)}$

$\qquad = \dfrac{\cos k\theta - i\sin k\theta}{\cos^2 k\theta + i\sin^2 k\theta} = \dfrac{\cos k\theta - i\sin k\theta}{1}$

$\qquad = \cos k\theta - i\sin k\theta.$

(b) From $z^k = \cos k\theta + i\sin k\theta$ and $z^{-k} = \cos k\theta - i\sin k\theta$,

adding gives $z^k + z^{-k} = 2\cos k\theta$,

$\therefore \cos k\theta = \frac{1}{2}(z^k + z^{-k})$.

Subtracting gives $z^k - z^{-k} = 2i\sin k\theta$,

$\therefore \sin k\theta = \frac{1}{2i}(z^k - z^{-k})$.

$\cos 2\theta \sin^2\theta = \frac{1}{2}(z^2 + z^{-2})\frac{1}{(2i)^2}(z - z^{-1})^2$

$\qquad\qquad = -\frac{1}{8}\left(z^2 + \frac{1}{z^2}\right)\left(z - \frac{1}{z}\right)^2$

(c) Expanding this last expression gives

$-\frac{1}{8}(z^2 + z^{-2})(z^2 - 2 + z^{-2})$

$= -\frac{1}{8}(z^4 + 1 - 2z^2 - 2z^{-2} + 1 + z^{-4})$

$= -\frac{1}{8}(z^4 + z^{-4} + 2 - 2z^2 - 2z^{-2})$

$= -\frac{1}{8}(2\cos 4\theta + 2 - 4\cos 2\theta)$

$= \frac{1}{2}\cos 2\theta - \frac{1}{4}\cos 4\theta - \frac{1}{4}$,

so $a = -\frac{1}{4}$, $b = \frac{1}{2}$, $c = -\frac{1}{4}$.

3 (a) $(\cos x + i\sin x)^3$

$= \cos^3 x + 3\cos^2 x\,(i\sin x) + 3\cos x\,(i\sin x)^2 + (i\sin x)^3$

$= \cos^3 x + i\,3\cos^2 x\sin x - 3\cos x\sin^2 x - i\sin^3 x$

(b) $(\cos x + i\sin x)^3 = \cos 3x + i\sin 3x$

(c) Equating real parts:

$\cos 3x = \cos^3 x - 3\cos x\sin^2 x$

$= \cos^3 x - 3\cos x\,(1 - \cos^2 x)$

$= \cos^3 x - 3\cos x + 3\cos^3 x$

$= 4\cos^3 x - 3\cos x$

ANSWERS (Contd.)

8. SEQUENCES AND SERIES

Arithmetic sequences – pp. 60-61

1 Let the first term be a and the common difference be d:

$a = 3$ and $u_{10} = a + 9d = 3 + 9d = -15 \Rightarrow d = -2$.

$S_{100} = \frac{n}{2}(2a + (n-1)d) = \frac{100}{2}\{2 \times 3 + 99 \times (-2)\} = -9600$.

or

$S_{100} = an + \frac{n}{2}(n-1)d = 300 + 50 \times 99 \times -2 = -9600$

2 $\displaystyle\sum_{r=1}^{n}(6 + 2r) = \sum_{r=1}^{n}6 + 2\sum_{r=1}^{n}r$

$\qquad = 6n + 2\left[\frac{1}{2}n(n+1)\right] = 7n + n^2$

$\displaystyle\sum_{r=1}^{3k}(6 + 2r) = 7(3k) + (3k)^2 = 21k + 9k^2$.

$\displaystyle\sum_{r=k+1}^{3k}(6 + 2r) = \sum_{r=1}^{3k}(6 + 2r) - \sum_{r=1}^{k}(6 + 2r)$

$\qquad = 21k + 9k^2 - 7k - k^2$

$\qquad = 14k + 8k^2$

3 (a) $u_1 = S_1 = 2 \times 1 - 1^2 = 1$

$\qquad u_2 = S_2 - S_1 = 2 \times 2 - 2^2 - 1 = -1$

$\qquad u_3 = S_3 - S_2 = 2 \times 3 - 3^2 - (4 - 4) = -3$

$\quad$ (b) $u_n = S_n - S_{n-1}$

$\qquad = 2n - n^2 - (2(n-1) - (n-1)^2)$

$\qquad = 2n - n^2 - (2n - 2 - n^2 + 2n - 1)$

$\qquad = 2n - n^2 - (-n^2 + 4n - 3)$

$\qquad = 2n - n^2 + n^2 - 4n + 3$

$\qquad = -2n + 3$

Geometric sequences – pp. 62-63

1 If the series is $a + ar + ar^2 + ar^3 + \ldots$

then $ar^2 = \frac{5}{4}$, $ar^7 = \frac{5}{128}$

$\dfrac{ar^7}{ar^2} = \dfrac{\frac{5}{128}}{\frac{5}{4}}$

$\Rightarrow r^5 = \frac{1}{32}, r = \frac{1}{2}$

$a\left(\frac{1}{2}\right)^2 = \frac{5}{4} \Rightarrow a = 5$

$S_8 = \dfrac{5\left(1 - \left(\frac{1}{2}\right)^7\right)}{1 - \frac{1}{2}} = 10 \times \frac{127}{128}$

$\qquad = \frac{635}{64}$

2 If the first term is a and the common ratio is r then

$ar = -2$, $ar^3 = -\frac{2}{9}$

$\therefore \dfrac{ar^3}{ar} = \dfrac{-\frac{2}{9}}{-2} = \frac{1}{9}$

$\therefore r^2 = \frac{1}{9} \Rightarrow r = \pm\frac{1}{3}$.

As both satisfy $|r| < 1$, each gives a sum to infinity.

$r = \frac{1}{3}$ gives $a = -6$ and sum to infinity $\dfrac{-6}{1 - \frac{1}{3}} = -9$.

$r = -\frac{1}{3}$ gives $a = 6$ and sum to infinity $\dfrac{6}{1 - \left(-\frac{1}{3}\right)} = \dfrac{6}{\frac{4}{3}} = \dfrac{9}{2}$

3 AP: $4, 4 + d, 4 + 2d$

$\therefore S_3 = 12 + 3d$

GP: $4, 4r, 4r^2$

$\therefore S_3 = 4 + 4r + 4r^2$

If $r = 2$ then GP is $4, 8, 16$ and $S_3 = 28$

$\therefore 4 + 4r + 4r^2 = 28$

$\therefore r^2 + r + 1 = 7$

$\therefore r^2 + r - 6 = 0$

$\therefore r = -3, 2$

$\therefore$ alternative $r = -3$,

$12 + 3d = 28$

$\quad \therefore 3d = 16$

$\quad \therefore d = \frac{16}{3}$

4 $S_n = \frac{n}{2}(2a + (n-1)d)$

$\therefore S_7 = \frac{7}{2}(2a + 6d) = 7a + 21d = 66\frac{1}{2}$ $\quad$ ☐1

$u_n = a + (n-1)d$

$\therefore u_7 = a + 6d = 16$ $\quad$ ☐2

☐2 $\times 7$ $\quad 7a + 42d = 112$ $\quad$ ☐3

$\qquad\qquad 7a + 21d = 66\frac{1}{2}$ $\quad$ ☐1

☐3 $-$ ☐1 $\qquad 21d = 45\frac{1}{2}$

$\qquad\qquad \therefore d = 2\frac{1}{6} \rightarrow$ $\quad$ ☐1

$\qquad 7a + 21 \times 2\frac{1}{6} = 66\frac{1}{2}$

$\qquad \therefore 7a + 45\frac{1}{2} = 66\frac{1}{2}$

$\qquad\qquad \therefore 7a = 21$

$\qquad\qquad \therefore a = 3$

$\therefore u_3 = a + 2d = 3 + 2 \times 2\frac{1}{6} = 7\frac{1}{3}$

contd

MacLaurin series – pp 64–65

1 $f(x) = \tan x, \quad f(0) = 0$

$f'(x) = \sec^2 x = (\cos x)^{-2}, \quad f'(0) = 1$

$f''(x) = -2(\cos x)^{-3}\cdot-\sin x$

$\quad = 2\sin x \cos^{-3} x, \quad f''(0) = 0$

$f'''(x) = 2\sin x\,(-3\cos^{-4}x)(-\sin x) + 2\cos x \cos^{-3}x$

$f'''(0) = 2$

$f^{iv}(x) = 6\sin^2 x(-4\cos^{-5}x) + \cos^{-4}x\,(12\sin x \cos x)$
$\qquad -4\cos^{-3}x\cdot(-\sin x), \quad f^{iv}(0) = 0$

$\therefore \tan x = x + \dfrac{2x^3}{3!} + \ldots$

$\qquad = x + \dfrac{1}{3}x^3 + \ldots$

2 $f(x) = \ln(x+2), \qquad f(0) = \ln 2$

$f'(x) = \dfrac{1}{x+2}, \qquad f'(0) = \dfrac{1}{2}$

$f''(x) = -(x+2)^{-2}, \qquad f''(0) = \dfrac{1}{4}$

$\therefore \ln(x+2) = \ln 2 + \dfrac{1}{2}x - \dfrac{\frac{1}{4}x^2}{2!} + \ldots$

$\qquad = \ln 2 + \dfrac{1}{2}x - \dfrac{1}{8}x^2 + \ldots$

Alternative method

$\ln(x+2) = \ln 2\left(1 + \dfrac{x}{2}\right)$

$\qquad = \ln 2 + \ln\left(1 + \dfrac{x}{2}\right)$

$\qquad\qquad$ valid for $-1 < \dfrac{x}{2} \leqslant 1 \Rightarrow -2 < x \leqslant 2$

$\qquad = \ln 2 + \dfrac{x}{2} - \dfrac{\left(\frac{x}{2}\right)^2}{2}$ [from result on table on
$\qquad\qquad\qquad\qquad\qquad\qquad$ page 65]

$\qquad = \ln 2 + \dfrac{x}{2} - \dfrac{x^2}{8}$

3 $\sin x = x - \dfrac{x^3}{3!} + \dfrac{x^5}{5!} - \ldots$

$\qquad = x - \dfrac{x^3}{6} + \dfrac{x^5}{120} - \ldots$

so, replacing x by $3x$,

$\sin 3x = 3x - \dfrac{27x^3}{6} + \dfrac{243x^5}{120} - \ldots$

$\qquad = 3x - \dfrac{9x^3}{2} + \dfrac{81x^5}{40} - \ldots$

$(1+x)\sin 3x = (1+x)\left(3x - \dfrac{9x^3}{2} + \dfrac{81x^5}{40} - \ldots\right)$

$\qquad = 3x - \dfrac{9x^3}{2} + \dfrac{81x^5}{40} - \ldots + 3x^2 - \dfrac{9x^4}{2} + \ldots$

$\qquad = 3x + 3x^2 - \dfrac{9}{2}x^3 - \dfrac{9}{2}x^4 + \dfrac{81}{40}x^5 + \ldots$

Associated series – pp. 66–67

Page 66

1 Let $f(x) = \ln(3-x): \quad f(0) = \ln 3;$

$f'(x) = -\dfrac{1}{3-x}, \qquad f'(0) = -\dfrac{1}{3}$

$f''(x) = -\dfrac{1}{(3-x)^2}, \qquad f''(0) = -\dfrac{1}{9}$

So $f(x) = \ln 3 - \dfrac{x}{3} - \dfrac{1}{2}\dfrac{x^2}{9} + \ldots = \ln 3 - \dfrac{x}{3} - \dfrac{x^2}{18} + \ldots$

$x^2 \ln(3-x) = x^2 \ln 3 - \dfrac{x^3}{3} - \dfrac{x^4}{18} + \ldots$

and $x^2 \ln(3+x) = x^2 \ln 3 + \dfrac{x^3}{3} - \dfrac{x^4}{18} + \ldots$

$x^2 \ln(9-x^2) = x^2 \ln(3-x) + x^2 \ln(3+x) = 2x^2 \ln 3 - \dfrac{x^4}{9} + \ldots$

2 $f(x) = e^{2x}, \qquad f(0) = 1$

$f'(x) = 2e^{2x}, \qquad f'(0) = 2$

$f''(x) = 4e^{2x}, \qquad f''(0) = 4$

$f'''(x) = 8e^{2x}, \qquad f'''(0) = 8$

$\therefore e^{2x} = 1 + 2x + \dfrac{4x^2}{2!} + \dfrac{8x^3}{3!}$

$\qquad = 1 + 2x + 2x^2 + \dfrac{4}{3}x^3$

$f(x) = \cos 3x \qquad\qquad \therefore f(0) = 1$

$f'(x) = -3\sin 3x \qquad \therefore f'(0) = 0$

$f''(x) = -9\cos 3x \qquad \therefore f''(0) = -9$

$f'''(x) = 27\sin 3x \qquad \therefore f'''(0) = 0$

$\therefore \cos 3x = 1 - \dfrac{9}{2}x^2$

$\therefore e^{2x}\cos 3x = \left(1 + 2x + 2x^2 + \dfrac{4}{3}x^3\right)\left(1 - \dfrac{9}{2}x^2\right)$

$\qquad = 1 + 2x + 2x^2 + \dfrac{4}{3}x^3 - \dfrac{9}{2}x^2 - 9x^3$

$\qquad = 1 + 2x - \dfrac{5}{2}x^2 - \dfrac{23}{3}x^3$

Page 67

1 $\dfrac{6}{(1+2x)(1-4x)} = 6(1 - 2x - 8x^2)^{-1}$

$\qquad\qquad\qquad\qquad = 6(1+y)^{-1}$ where $y = -2x - 8x^2$

$6(1+y)^{-1} = 6[1 - y + y^2 - y^3 + \ldots]$

$\qquad = 6[1 - (-2x - 8x^2) + (-2x - 8x^2)^2 - (-2x$
$\qquad\qquad - 8x^2)^3 + \ldots]$

$\qquad = 6[1 + (2x + 8x^2) + (4x^2 + 32x^3 + 64x^4)$
$\qquad\qquad - (-8x^3) \ldots]$

contd

ANSWERS (Contd.)

All other terms will involve powers of x^4 or higher.

$$= 6[1 + 2x + 8x^2 + 4x^2 + 32x^3 + 8x^3]$$

$$= 6 + 12x + 72x^2 + 240x^3$$

Valid for $\quad -1 < 2x < 1$ and $-1 < 4x < 1$

So $\qquad \frac{1}{2} > x > -\frac{1}{2} \quad -\frac{1}{4} < x < \frac{1}{4}$

$$-\frac{1}{4} < x < \frac{1}{4}$$

The series is only valid when both conditions apply,
so $-\frac{1}{4} < x < \frac{1}{4}$.

2 $\quad \dfrac{x^2 + 8x + 11}{(x + 1)(x + 3)^2} = \dfrac{A}{x + 1} + \dfrac{B}{x + 3} + \dfrac{C}{(x + 3)^2}$

$x^2 + 8x + 11 = A(x + 3)^2 + B(x + 1)(x + 3) + C(x + 1)$

$x = -1$ gives $4 = 4A$, $A = 1$;

$x = -3$ gives $-4 = -2C$, $C = 2$;

$x = -2$ gives $-1 = A - B - C$, $-1 = 1 - B + 2$, $\quad B = 0$;

or equating the coefficient of x^2 on both sides:

$1 = A + B$, so $B = 0$.

$f(x) = \dfrac{x^2 + 8x + 11}{(x + 1)(x + 3)^2} = \dfrac{1}{x + 1} + \dfrac{2}{(x + 3)^2}$, $\quad f(0) = \dfrac{11}{9}$

$f'(x) = -\dfrac{1}{(x + 1)^2} - \dfrac{4}{(x + 3)^3}$, $\qquad f'(0) = \dfrac{-31}{27}$

$f''(x) = \dfrac{2}{(x + 1)^3} + \dfrac{12}{(x + 3)^4}$, $\qquad f''(0) = \dfrac{58}{27}$

$f(x) = \dfrac{11}{9} - \dfrac{31}{27}x + \dfrac{58}{27}\dfrac{x^2}{2} - \ldots = \dfrac{11}{9} - \dfrac{31}{27}x + \dfrac{29}{27}x^2 + \ldots$

The series for $\dfrac{1}{1 + x}$ converges for $|x| < 1$.

Writing $\dfrac{1}{x + 3} = \dfrac{1}{3}\left(1 + \dfrac{x}{3}\right)^{-1}$ shows that the series for $\dfrac{1}{3 + x}$ converges for $|x| < 3$.

For both terms to converge, we need $|x| < 1$.

9. VECTORS

Basic skills information – pp. 68-69

1 $\quad v \times w = \begin{vmatrix} i & j & k \\ 2 & -1 & 2 \\ -1 & 0 & 2 \end{vmatrix}$

$$= i\begin{vmatrix} -1 & 2 \\ 0 & 2 \end{vmatrix} - j\begin{vmatrix} 2 & 2 \\ -1 & 2 \end{vmatrix} + k\begin{vmatrix} 2 & -1 \\ -1 & 0 \end{vmatrix}$$

$$= -2i - 6j - k$$

$(i + j - k) \cdot (-2i - 6j - k) = 1(-2) + 1(-6) + (-1)(-1) = -7.$

2 $\quad a \times b = \begin{vmatrix} i & j & k \\ 2 & -1 & 3 \\ -1 & 0 & 1 \end{vmatrix}$

$$= i\begin{vmatrix} -1 & 3 \\ 0 & 1 \end{vmatrix} - j\begin{vmatrix} 2 & 3 \\ -1 & 1 \end{vmatrix} + k\begin{vmatrix} 2 & -1 \\ -1 & 0 \end{vmatrix}$$

$$= -i - 5j + k$$

$(a \times b)$ perpendicular to c then $c \cdot a \times b = 0$

$$\therefore \begin{bmatrix} -1 \\ -5 \\ -1 \end{bmatrix} \cdot \begin{bmatrix} -2 \\ t \\ -1 \end{bmatrix} = 0$$

$\therefore 2 - 5t + 1 = 0$, so $t = \dfrac{3}{5}$

Equations of lines – pp. 70-71

1 (a) Parametric equations for the lines are:

L_1: $x = -3 - 2s$, $y = 1 + s$, $z = 5 + 3s$

L_2: $x = 1 + t$, $y = 1 - t$, $z = 3 - t$

Equating the corresponding x- and y-coordinates and solving for s and t gives

x: $-3 - 2s = 1 + t$, y: $1 + s = 1 - t$, $\Rightarrow s = -4$, $t = 4$.

For L_1 this gives $z = -7$, and for L_2 $z = -1$. Hence the lines do not intersect.

(b) A vector perpendicular to the directions of L_1 and L_2 is given by

$$\begin{vmatrix} i & j & k \\ -2 & 1 & 3 \\ 1 & -1 & -1 \end{vmatrix} = 2i + j + k,$$

so L_3 has parametric equations

$x = -1 + 2u$, $y = u$, $z = 2 + u$.

(c) The lines L_3 and L_2 intersect if the equations

$-1 + 2u = 1 + t$, $u = 1 - t$, $2 + u = 3 - t$ have a unique solution.

Solving the first two equations gives $u = 1$, $t = 0$, and these satisfy the third equation also, so the lines intersect.

Q is the point $(1, 1, 3)$ ($t = 0$ for L_2 or $u = 1$ for L_3)

(d) $P(-1, 0, 2)$: for L_1 $x = -3 - 2s$. Setting $x = -1$ gives

$-1 = -3 - 2s \Rightarrow s = -1$.

$y = 1 + s \Rightarrow y = 0$, $z = 5 + 3s \Rightarrow z = 2$, which shows that P lies on L_1.

P lies on L_1, Q lies on L_2, and the line PQ is perpendicular to both lines.

contd

Equations of planes 1 – pp. 72-73

1 $\overrightarrow{AB} \times \overrightarrow{AC}$ gives $\begin{vmatrix} i & j & k \\ 1 & 1 & 2 \\ 0 & 2 & -1 \end{vmatrix} = i(-1 - 4) - j(-1) + k(2)$

$= -5i + j + 2k$

which is a vector normal to the plane.

$n = \begin{bmatrix} -5 \\ 1 \\ 2 \end{bmatrix}$

and the equation of the plane is $-5x + y + 2z = k$

where $k = 1 \times -5 + 2 \times 1 + -1 \times 2 = -5$ (using point A).

This gives the **Cartesian equation** $-5x + y + 2z = -5$.

Equations of planes 2 – pp. 74-75

1 $\overrightarrow{AB} = \begin{bmatrix} 1 \\ 1 \\ 2 \end{bmatrix}$, $\overrightarrow{AC} = \begin{bmatrix} 0 \\ 2 \\ -1 \end{bmatrix}$

We can use any of the points given to get the **vector equation**, for example:

$r = \begin{bmatrix} 1 \\ 2 \\ -1 \end{bmatrix} + s \begin{bmatrix} 1 \\ 1 \\ 2 \end{bmatrix} + t \begin{bmatrix} 0 \\ 2 \\ -1 \end{bmatrix}$

Intersections of lines and planes – pp. 78-79

1 (a) A normal vector to the plane is given by the direction vector of the line: $n = i + 2j + k$

$n = \begin{bmatrix} 1 \\ 2 \\ 1 \end{bmatrix}$, $u = \begin{bmatrix} 2 \\ 1 \\ -1 \end{bmatrix}$, $n.u = \begin{bmatrix} 1 \\ 2 \\ 1 \end{bmatrix} . \begin{bmatrix} 2 \\ 1 \\ -1 \end{bmatrix} = 2 + 2 - 1 = 3$

An equation for the plane is $x + 2y + z = 3$.

or

An equation for the plane is

$1.(x - 2) + 2.(y - 1) + 1.(z + 1) = 0$, i.e. $x + 2y + z = 3$.

Parametric equations for L are:

$x = \lambda - 3, y = 2\lambda, z = \lambda + 12$

(b) The plane and L meet when

$(\lambda - 3) + 2(2\lambda) + (\lambda + 12) = 3 \Rightarrow 6\lambda + 9 = 3 \Rightarrow \lambda = -1$.

Setting $\lambda = -1$ in the parametric equations for L gives $B(-4, -2, 11)$.

(c) $AB^2 = (2 + 4)^2 + (1 + 2)^2 + (-1 - 11)^2 = 1$,

so $AB = \sqrt{189} = 3\sqrt{21}$.

This gives the shortest distance from A to L because B is the foot of the perpendicular from A onto L.

Angles between lines and planes – pp. 80-81

1 (a) Interchange the first two equations:

R_1	1	2	1	3	
R_2	2	-1	1	8	
R_3	-1	3	2	1	
R_1	1	2	1	3	
R_4	0	-5	-1	2	$R_2 - 2R_1$
R_5	0	5	3	4	$R_3 + R_1$
R_1	1	2	1	3	
R_4	0	5	1	-2	$-R_4$
R_5	0	0	2	6	$R_4 + R_5$

Solving gives $z = 3, y = -1, x = 2$.

(b) Setting $y = -t$ in the two equations gives

$2x + t + z = 8$ ①

$x - 2t + z = 3$ ②

Solving ① and ② simultaneously gives

$x = 5 - 3t, z = 5t - 2$

Thus, $x = 5 - 3t, y = -t, z = 5t - 2$.

(c) We need the acute angle between the line and the plane.

Firstly, we find the acute angle between

$d = \begin{bmatrix} -3 \\ -1 \\ 5 \end{bmatrix}$,

the direction vector of the line, and

$n = \begin{bmatrix} -1 \\ 3 \\ 2 \end{bmatrix}$, the normal vector to the plane.

$d.n = (-3)(-1) + (-1)3 + (5)(2) = 10, |d| = \sqrt{35}, |n| = \sqrt{14}$.

$\cos\theta = \frac{10}{\sqrt{35} \sqrt{14}}$, giving $\theta = 63.1°$

to three significant figures.

Thus the angle between the line and the plane is $90° - 63.1° = 26.9°$

2 (a) $\overrightarrow{AB} = -4i - 12j$ $\overrightarrow{AC} = -i + j + 2k$

$\overrightarrow{AB} \times \overrightarrow{AC} = \begin{vmatrix} i & j & k \\ -4 & -12 & 0 \\ -1 & 1 & 2 \end{vmatrix} = -24i + 8j - 16k = 8\begin{bmatrix} -3 \\ 1 \\ -2 \end{bmatrix}$

$\therefore \pi_1: -3x + y - 2z = d$

Using $A(3, 4, 4)$: $-3 \times 3 + 4 - 2 \times 4 = d, d = -13$

$\therefore \pi_1: -3x + y - 2z = -13$

(or $3x - y + 2z = 13$).

contd

ANSWERS (Contd.)

(b)
$$\begin{array}{ccc|c} 3 & -1 & 2 & 13 \\ 3 & 1 & 3 & 25 \\ 3 & -1 & -1 & -5 \end{array} \quad \begin{array}{l} R_2 \to R_2 - R_1 \\ R_3 \to R_1 - R_3 \end{array}$$

$$\begin{array}{ccc|c} 3 & -1 & 2 & 13 \\ 0 & 2 & 1 & 12 \\ 0 & 0 & 3 & 18 \end{array}$$

$\therefore z = 6, \; 2y + 6 = 12, \quad y = 3$

$\therefore 3x - 3 + 12 = 13$

$\therefore x = \frac{4}{3}$

$\therefore$ Planes meet at a **single point** $\left(\frac{4}{3}, 3, 6\right)$

(c) Angle between the planes = angle between the normals.

$$n_2 = \begin{bmatrix} 3 \\ 1 \\ 3 \end{bmatrix} \qquad n_3 = \begin{bmatrix} 3 \\ -1 \\ -1 \end{bmatrix}$$

$\cos \theta = \frac{n_2 \cdot n_3}{|n_2||n_3|} = \frac{5}{\sqrt{19}\,\sqrt{11}}$

$\therefore \theta = 69 \cdot 8°$ (1 d.p.)

3 (a) $\overrightarrow{AB} = \begin{bmatrix} -2 \\ 4 \\ 4 \end{bmatrix} \qquad \overrightarrow{AC} = \begin{bmatrix} 1 \\ -5 \\ -2 \end{bmatrix}$

$\therefore \overrightarrow{AB} \times \overrightarrow{AC} = \begin{vmatrix} i & j & k \\ -2 & 4 & 4 \\ 1 & -5 & -2 \end{vmatrix} = 12i - j(4 - 4) + k(10 - 4)$

$= 12i + 6k = 6(2i + k)$

$\therefore \pi_1: \quad 2x + z = d$

$(1, 2, 0): \quad 2 \times 1 + 0 = d \quad \therefore d = 2$

$\therefore$ Equation is $2x + z = 2$

(b) Normal to plane $\pi_1 = n_1 = \begin{bmatrix} 2 \\ 0 \\ 1 \end{bmatrix}$;

direction vector of line $= d = \begin{bmatrix} 3 \\ 2 \\ -1 \end{bmatrix}$

$\therefore \theta$ = angle between n_1 and d:

$\cos \theta = \frac{n_1 \cdot d}{|n_1||d|} = \frac{5}{\sqrt{5}\,\sqrt{14}} \quad \therefore \theta = 53 \cdot 3°$

$\therefore$ Acute angle between line and plane = $36 \cdot 7°$ (1 d.p.)

10. DIFFERENTIAL EQUATIONS

First-order equations 1 - pp. 82-83

1 $\int \frac{dx}{x(1000 - x)} = \int k\,dt \quad *$

$\therefore \frac{A}{x} + \frac{B}{(1000 - x)} = \frac{A(1000 - x) + Bx}{x(1000 - x)}$

$\therefore A(1000 - x) + Bx = 1$

If $x = 0$ then $A = \frac{1}{1000}$

If $x = 1000$ then $B = \frac{1}{1000}$

$\therefore *$ becomes $\int \left(\frac{1}{1000x} + \frac{1}{1000(1000 - x)}\right) dx = \int k\,dt$

$\therefore \frac{1}{1000}(\ln x - \ln(1000 - x)) = kt + c$

$\therefore \ln \left(\frac{x}{1000 - x}\right) = 1000kt + 1000c$

$\therefore \frac{x}{1000 - x} = Ae^{1000kt}$

$\therefore x = 1000Ae^{1000kt} - xAe^{1000kt}$

$\therefore x\left(1 + Ae^{1000kt}\right) = 1000Ae^{1000kt}$

$\therefore x = \frac{1000\,Ae^{1000kt}}{1 + Ae^{1000kt}}$ ①

Substituting $t = 0, \; x = 0$ in ① gives

$1 = \frac{1000A}{1 + A}$

$\therefore 1 + A = 1000A$

$\therefore 1 = 999A$

$\therefore \frac{1}{999} = A$

Using ① when $x = 50$ and $t = 4$

$50 = \frac{\frac{1000}{999}e^{4000k}}{1 + \frac{1}{999}e^{4000k}}$

$\therefore 50 = \frac{1000\,e^{4000k}}{999 + e^{4000k}}$

$\therefore 50(999 + e^{4000k}) = 1000e^{4000k}$

$\therefore 999 + e^{4000k} = 20e^{4000k}$

$\therefore 999 = 19e^{4000k}$

$\therefore e^{4000k} = \frac{999}{19}$

$\therefore 4000k = \ln \left(\frac{999}{19}\right)$

$\therefore k = \frac{\ln\left(\frac{999}{19}\right)}{4000} = 0 \cdot 0009906$

Thus ① becomes $x = \frac{1000\,e^{.9906t}}{999 + e^{.9906t}}$

$\therefore$ When $t = 6, \quad x = \frac{1000\,e^{.9906 \times 6}}{999 + e^{.9906 \times 6}} = 276 \cdot 22$

so 276 people infected after six days.

First-order equations 2 - pp. 84-85

1 $\frac{dy}{dx} - \frac{2}{x}y = x^2 e^{2x}$

integrating factor is $e = e^{\int \frac{-2}{x}dx} = e^{-2\ln x} = e^{\ln x^{-2}} = x^{-2} = \frac{1}{x^2}$

So $\frac{1}{x^2}\frac{dy}{dx} - \frac{2}{x^3}y = e^{2x}$

$\Rightarrow \frac{d}{dx}\left(\frac{y}{x^2}\right) = e^{2x}$

$\Rightarrow \frac{y}{x^2} = \int e^{2x}dx$

$\Rightarrow \frac{y}{x^2} = \frac{1}{2}e^{2x} + c$

contd

so $y = \frac{1}{2}x^2 e^{2x} + cx^2$.

$y(1) = 0$ gives $0 = \frac{1}{2}e^2 + c$, so $c = -\frac{1}{2}e^2$

and $y = \frac{1}{2}x^2 e^{2x} - \frac{1}{2}e^2 x^2 \Rightarrow y = \frac{1}{2}x^2(e^{2x} - e^2)$.

2 (a) $\frac{dy}{dx} - \frac{3}{x}y + \frac{1}{x^2} = 0$

$\therefore \frac{dy}{dx} - \frac{3}{x}y = -\frac{1}{x^2}$

$e^{\int -\frac{3}{x}dx} = e^{-3\ln x} = e^{\ln x^{-3}} = x^{-3} = \frac{1}{x^3}$

$\therefore \frac{1}{x^3}\frac{dy}{dx} - \frac{3}{x^4}y = -\frac{1}{x^5}dx$

to give $\frac{1}{x^3}y = \int -\frac{1}{x^5}dx$

$\therefore \frac{1}{x^3}y = +\frac{1}{4x^4} + c$

$\therefore y = +\frac{1}{4x} + cx^3$

$\left(\frac{1}{2}, 2\right): 2 = \frac{1}{4} \times \frac{1}{2} + c \times \frac{1}{8}$

$\therefore 2 = \frac{1}{8} + \frac{1}{8}c$

$\therefore \frac{15}{8} = \frac{1}{8}c$

$\therefore 15 = c$

$\therefore y = \frac{1}{4x} + 15x^3$

(b) $\frac{1}{y}\frac{dy}{dx} - 3x = 1$

$\therefore \int \left(\frac{1}{y}dy\right) = \int (1 + 3x)\,dx$

$\therefore \ln y = x + \frac{3}{2}x^2 + c$

$\therefore y = e^{x + \frac{3}{2}x^2 + c} = Ae^{x + \frac{3}{2}x^2}$

$(0, 1): \quad 1 = Ae^0 \therefore A = 1$

$\therefore$ Particular solution is $y = e^{x + \frac{3}{2}x^2}$

(c) Old model: $y = \frac{1}{4 \times 1} + 15 \times 1^3 = 15\frac{1}{4}$

New model: $y = e^{\frac{5}{2}} = 12 \cdot 82$

Reduction of $15 \cdot 9\%$ (3 s.f.)

3 (a) $\int \frac{dP}{16k - P} = \int \frac{dt}{1000}$

$\therefore -\ln(16k - P) = \frac{t}{1000} + c$

$\therefore \ln(16k - P)^{-1} = \frac{t}{1000} + c$

$\therefore e^{\ln(16k - P)^{-1}} = e^{\frac{t}{1000} + c}$

$\therefore \frac{1}{16k - P} = Ae^{\frac{t}{1000}}$

$\therefore \frac{1}{Ae^{\frac{t}{1000}}} = 16k - P$

$\therefore P = 16k - \frac{1}{Ae^{\frac{t}{1000}}}$ QED

(b) $(0, 100) \rightarrow 100 = 16k - \frac{1}{A}$ ☐1

$(1000, 200) \rightarrow 200 = 16k - \frac{1}{Ae}$ ☐2

☐2 − ☐1 $\rightarrow 100 = \frac{1}{A} - \frac{1}{Ae}$

$\therefore 100Ae = e - 1$

$\therefore A = \frac{e - 1}{100e}$

$\therefore 100 = 16k - \frac{100e}{e - 1}$

$\therefore 16k = 100 + \frac{100e}{e - 1}$

$\therefore k = \frac{100}{16} + \frac{100e}{16(e - 1)}$

(c) Max P as $t \rightarrow \infty$

$\therefore P \rightarrow 16k = 16\left(\frac{100}{16} + \frac{100e}{16(e - 1)}\right)$

$= 100 + \frac{100e}{e - 1} = 258 \cdot 197...$

$\therefore$ Long-term max $= 258$

Second-order equations 1 – pp. 86–87

1 The auxiliary equation is

$m^2 + 6m + 9 = 0$

$(m + 3)^2 = 0$ giving repeated root $m = -3$.

The general solution is $y = (A + Bx)e^{-3x}$;

$y = 2$ when $x = 0$ gives $A = 2$.

$y = (2 + Bx)e^{-3x} \Rightarrow \frac{dy}{dx} = -3(2 + Bx)e^{-3x} + Be^{-3x}$,

so $\frac{dy}{dx} = -3$ when $x = 0$ gives $B = 3$.

The solution is $y = (2 + 3x)e^{-3x}$.

2 $\frac{d^2y}{dx^2} + 4\frac{dy}{dx} - 5y = 0$

$m^2 + 4m - 5 = 0$

$(m + 5)(m - 1) = 0$

$m = -5 \qquad m = 1$

$y = Ae^{-5x} + Be^x$

$\frac{dy}{dx} = -5Ae^{-5x} + Be^x$

when $x = 0$, $y = 1$ gives $1 = A + B$

when $x = 0$, $\frac{dy}{dx} = 10$ gives $10 = -5A + B$

$\Rightarrow 6A = -9, A = -\frac{3}{2}, B = \frac{5}{2}$

so $y = -\frac{3}{2}e^{-5x} + \frac{5}{2}e^x$

3 Auxiliary equation is

$m^2 + 2m + 5 = 0 \Rightarrow (m + 1)^2 + 4 = 0 \Rightarrow m = -1 \pm 2i$.

The general solution is $y = e^{-x}(A\cos 2x + B\sin 2x)$.

$y = 1$ when $x = 0$ gives $A = 1$;

$y = 2$ when $x = \frac{\pi}{4}$ gives $2 = e^{-\frac{\pi}{4}}(0 + B) \Rightarrow B = 2e^{\frac{\pi}{4}}$.

The solution is $y = e^{-x}(\cos 2x + 2e^{\frac{\pi}{4}}\sin 2x)$.

Second-order equations 2 – pp. 88–89

1 Auxiliary equation is
$m^2 + 4m - 5 = 0 \Rightarrow (m - 1)(m + 5) = 0$.

Roots are $m = 1$, $m = -5$; complementary function:
$y = Ae^x + Be^{-5x}$.

For a particular solution try $y = Ce^{2x} + D$;

$\frac{dy}{dx} = 2Ce^{2x}, \quad \frac{d^2y}{dx^2} = 4Ce^{2x}$.

contd

ANSWERS (Contd.)

Substituting into the full equation gives

$$4Ce^{2x} + 8Ce^{2x} - 5Ce^{2x} - 5D = 7e^{2x} + 10$$

i.e. $7Ce^{2x} - 5D = 7e^{2x} + 10$; equating coefficients gives

$C = 1$ and $D = -2$.

The general solution of the full equation is

$$y = Ae^x + Be^{-5x} + e^{2x} - 2.$$

$\frac{dy}{dx} = Ae^x - 5Be^{-5x} + 2e^{2x}$

$y(0) = 1$ gives $A + B = 2$; $y'(0) = 10$ gives $A - 5B = 8$.

Solving gives $A = 3$, $B = -1$.

The particular solution is $y = 3e^x - e^{-5x} + e^{2x} - 2$.

2   Auxiliary equation is

$$m^2 + 4m + 8 = 0 \Rightarrow (m + 2)^2 + 4 = 0 \Rightarrow m = -2 \pm 2i.$$

The complementary function is

$$y = e^{-2x}(A\cos 2x + B\sin 2x).$$

For a particular solution try $y = Cx^2 + Dx + E$;

$y' = 2Cx + D, y'' = 2C$.

Substituting into the full equation:

$$2C + 4(2Cx + D) + 8(Cx^2 + Dx + E) = 8x^2 + 16x + 6$$

i.e. $8Cx^2 + (8C + 8D)x + 2C + 4D + 8E = 8x^2 + 16x + 6$.

Equating coefficients gives

$C = 1, 8 + 8D = 16, 6 + 8E = 6$, so $C = 1, D = 1, E = 0$.

The general solution of the full equation is

$$y = e^{-2x}(A\cos 2x + B\sin 2x) + x^2 + x.$$

11. NUMBER THEORY AND PROOF

Numbers, notation and direct proof - pp. 90-91

1   LHS $= \frac{1}{\cos^2 x}$

$= \frac{\sin^2 x + \cos^2 x}{\cos^2 x}$

$= \frac{\sin^2 x}{\cos^2 x} + \frac{\cos^2 x}{\cos^2 x}$

$= \tan^2 x + 1$

$=$ RHS as required

or

RHS $= 1 + \tan^2 x$

$= 1 + \frac{\sin^2 x}{\cos^2 x}$

$= \frac{\cos^2 x}{\cos^2 x} + \frac{\sin^2 x}{\cos^2 x}$

$= \frac{\cos^2 x + \sin^2 x}{\cos^2 x}$

$= \frac{1}{\cos^2 x}$

$= \sec^2 x$

$=$ LHS as required

2   Let the two positive integers be $m = 2p$, $n = 2q - 1$, where $p, q \in \mathbb{N}$.

Then $mn = 2p(2q - 1)$

$= 4pq - 2p$

$= 2(2pq - p)$

which is divisible by 2, so $2|mn$

$\therefore mn$ is even QED

Proof by contradiction and by contrapositive - pp. 92-93

1   Contrapositive is: $3 + x$ rational $\Rightarrow x$ rational.

Proof

$3 + x$ rational $\Rightarrow \exists p, q$:

$3 + x = \frac{p}{q}, p, q \in \mathbb{Z}$

$\Rightarrow x = \frac{p}{q} - 3$

$\Rightarrow x = \frac{p - 3q}{q}$

$p, q \in \mathbb{Z} \therefore p - 3q \in \mathbb{Z}$

$\therefore 3 + x$ rational $\Rightarrow x$ rational

$\therefore$ By contrapositive, x irrational $\Rightarrow 3 + x$ irrational.

Further proof - pp. 94-95

1   A: Let $p = 2m$ (even)

$q = 2n + 1$ (odd)

so $p^3 + q^2 = (2m)^3 + (2n + 1)^2 = 8m^3 + 4n^2 + 4n + 1$

$= 2(4m^3 + 2n^2 + 2n) + 1$.

This is odd, being of the form $2k + 1$, where k is an integer, so A is true.

B: The counterexample $m = 3$ shows that B is false.

m^2 is divisible by 9, but m is not divisible by 9

contd

2 'If' part:

$100a + 10b + c = 99a + 9b + a + b + c$,
so if $a + b + c = 9k$, where k is an integer,
$100a + 10b + c = 99a + 9b + 9k$, which is divisible by 9.

'Only if' part:

If $100a + 10b + c = 9m$, where m is an integer,
then $100a + 10b + c = 99a + 9b + a + b + c$ gives
$9m = 99a + 9b + a + b + c$,

and so $a + b + c = 9m - 99a - 9b$ is divisible by 9.

3 **A:** True $p^3 - p = p(p^2 - 1) = p(p + 1)(p - 1)$

$= (p - 1)\,p(p + 1)$

$(p - 1), p, (p + 1)$ are consecutive

∴ at least one of $(p - 1), p, (p + 1)$ is even

∴ $2|(p^3 - 1)$ $\boxed{1}$

And one of $(p - 1), p, (p + 1)$ is divisible by 3

∴ $3|(p^3 - p)$ $\boxed{2}$

From $\boxed{1}$ and $\boxed{2}$ 2 and $3|(p^3 - p)$

∴ $6|(p^3 - p)$ $\forall p \in \mathbb{N}, p > 2$

B: False

Counterexample $p = 5$,

$p^4 - p = 620$ and $\frac{620}{6} = 103\frac{1}{3}$

620 is not divisible by 6.

∴ false

4 **A:** True Seek n^3 even $\Rightarrow n$ even

∴ n odd $\Rightarrow n^3$ odd, using contrapositive

Proof: n odd $\Rightarrow n = 2p - 1$, $p \in \mathbb{N}$

$n^3 = (2p - 1)^3 = (4p^2 - 4p + 1)(2p - 1)$

$= 8p^3 - 4p^2 - 8p^2 + 4p + 2p - 1$

$= 8p^3 - 12p^2 + 2p - 1$

$= 2(4p^3 - 6p^2 + 3p) - 1$

∴ n^3 odd

∴ n odd $\Rightarrow n^3$ odd

∴ n not even $\Rightarrow n^3$ not even

∴ By contrapositive n^3 even $\Rightarrow n$ even

B: False   Counterexample $n = 2$

$2^3 = 8$ and $8|8$ ∴ n^3 a multiple of 8 but $8 > 2$
∴ n is not a multiple of 8.

Proof by induction - pp. 96-97

1 Prove by induction that $\sum_{r=1}^{n}(6r^2 + 4r) = n(n + 1)(2n + 3)$.

When $n = 1$

LHS: $6(1)^2 + 4(1) = 10$, RHS: $1 \times 2 \times 5 = 10$

so result true for $n = 1$.

Assume true for $n = k$

$\sum_{r=1}^{k}(6r^2 + 4r) = k(k + 1)(2k + 3)$.

Consider $n = k + 1$

Then $\sum_{r=1}^{k+1}(6r^2 + 4r) = \sum_{r=1}^{k}(6r^2 + 4r) + \left[6(k + 1)^2 + 4(k + 1)\right]$

$= k(k + 1)(2k + 3) + 6(k + 1)^2 + 4(k + 1)$

$= (k + 1)[k(2k + 3) + 6(k + 1) + 4]$

$= (k + 1)\left[2k^2 + 9k + 10\right]$

$= (k + 1)(k + 2)(2k + 5)$

$= (k + 1)((k + 1) + 1)(2(k + 1) + 3)$

which is the result for k with k replaced by $k + 1$.

So if true for $n = k$, then true for $n = k + 1$. But true for $n = 1$, then by induction, true for all integers $n \geq 1$.

2 Prove by induction on n that for $x > 0$

$(1 + x)^n \geq 1 + nx + \frac{1}{2}n(n - 1)x^2$

for all positive integers n.

When $n = 1$

LHS: $(1 + x)^1 = (1 + x)$, RHS: $1 + x + \frac{1}{2}1.(1 - 1)x^2 = 1 + x$

LHS $\geq$ RHS, so true for $n = 1$.

Assume true for $n = k$

$(1 + x)^k \geq 1 + kx + \frac{1}{2}k(k - 1)x^2$

Consider $n = k + 1$

$(1 + x)^{k+1} = (1 + x)(1 + x)^k \geq (1 + x)\left(1 + kx + \frac{1}{2}k(k - 1)x^2\right)$

because $1 + x > 0$.

So $(1 + x)^{k+1} \geq (1 + x)(1 + kx + \frac{1}{2}k(k - 1)x^2)$

$= 1 + kx + \frac{1}{2}k(k - 1)x^2 + x + kx^2 + \frac{1}{2}k(k - 1)x^3$

$= 1 + (k + 1)x + \left[\frac{1}{2}k(k - 1) + k\right]x^2 + \frac{1}{2}k(k - 1)x^3$

$\geq 1 + (k + 1)x + \left[\frac{1}{2}k(k - 1) + k\right]x^2$

since $k \geq 1$ and $x > 0 \Rightarrow \frac{1}{2}k(k - 1)x^3 \geq 0$

$\frac{1}{2}k(k - 1) + k = \frac{1}{2}\left[k(k - 1) + 2k\right]$

$= \frac{1}{2}(k + 1)k$

so $(1 + x)^{k+1} \geq 1 + (k + 1)x + \frac{1}{2}(k + 1)kx^2$

which is the result for k with k replaced by $k + 1$.
So if true for k, then true for $k + 1$. But true for $n = 1$,
hence by induction true for all integers $n \geq 1$.

ANSWERS (Contd.)

3 $(AB)^{-1} = B^{-1}A^{-1}$. ☐1

LHS: $(A^1)^{-1} = A^{-1}$, RHS: $(A^{-1})^1 = A^{-1}$, so the result is true for $n = 1$. ☐2

Assume that, for some integer k, $(A^k)^{-1} = (A^{-1})^k$ ☐3

and consider $(A^{k+1})^{-1}$.

$A^{k+1} = A^k A^1$

$(A^{k+1})^{-1} = (A^k A^1)^{-1}$

$= (A^1)^{-1}(A^k)^{-1}$ from ☐1

$= A^{-1}(A^{-1})^k$ from ☐2 and ☐3

$= (A^{-1})^{k+1}$

or

$[A^{k+1} = AB$, where $B = A^k$,

so $(A^{k+1})^{-1} = (AB)^{-1} = B^{-1}A^{-1} = (A^k)^{-1}A^{-1}$.

By hypothesis, $(A^k)^{-1} = (A^{-1})^k$,

so $(A^{k+1})^{-1} = (A^k)^{-1}A^{-1} = (A^{-1})^k A^{-1} = (A^{-1})^{k+1}]$.

This proves that the inverse of A^{k+1} is $(A^{-1})^{k+1}$.

So the truth for $n = k$ implies the truth for $n = k + 1$, and because the result holds for $n = 1$, the result, by induction, is true for all integers $n \geqslant 1$.

4 LHS: $\frac{d}{dx}(xe^x) = 1.e^x + xe^x = (x + 1)e^x$

RHS: $(x + 1)e^x$

hence the result holds for $n = 1$.

Assume that, for some positive integer k,

$\frac{d^k}{dx^k}(xe^x) = (x + k)e^x$.

Then $\frac{d^{k+1}}{dx^{k+1}}(xe^x) = \frac{d}{dx}\left(\frac{d^k}{dx^k}(xe^x)\right) = \frac{d}{dx}((x + k)e^x)$ by above assumption.

$\frac{d}{dx}((x + k)e^x) = e^x + (x + k)e^x = (x + (k + 1))e^x$

So, if the result is true for $n = k$, it is true for $n = k + 1$. But the result is true for $n = 1$, hence by induction true for all integers $n \geqslant 1$.

5 $n = 1$ LHS $= 1 \times 2 = 2$

RHS $= \frac{1}{3} \times 1 \times 2 \times 3 = 2$

$\therefore$ LHS = RHS $\therefore$ true for $n = 1$

Assume true for some $n = k$

i.e. $\sum_{r=1}^{k} r(r + 1) = \frac{1}{3}k(k + 1)(k + 2)$

Consider $n = k + 1$.

$\sum_{r=1}^{k+1} r(r + 1) = \sum_{r=1}^{k} r(r + 1) + (k + 1)(k + 2)$

$= \frac{1}{3}k(k + 1)(k + 2) + (k + 1)(k + 2)$

$= \frac{1}{3}(k + 1)(k + 2)[k + 3]$

$= \frac{1}{3}(k + 1)((k + 1) + 1)((k + 1) + 2)$

$\therefore$ if true for $n = k$ then true for $n = k + 1$, and since true for $n = 1$, by induction, true for all $n \in \mathbb{N}$.

6 For $n = 3$ $7^n - 4^{n-2} = 7^3 - 4^1 = 343 - 4 = 339$

and $339 \div 3 = 113$

$\therefore 3|(7^n - 4^{n-2})$ when $n = 3$.

Assume true for some $n = k$.

Then $3|(7^k - 4^{k-2})$ *

Consider $n = k + 1$.

$7^{k+1} - 4^{k-1} = 7 \times 7^k - 4 \times 4^{k-2}$

$= (6 + 1)7^k - (3 + 1)4^{k-2}$

$= 6 \times 7^k + 7^k - 3 \times 4^{k-2} - 4^{k-2}$

$= 3(2 \times 7^k - 4^{k-2}) + 7^k - 4^{k-2}$

And since from * above $3|(7^k - 4^{k-2})$, $3|(7^{k+1} - 4^{k-1})$

If true for $n = k$ then true for $n = k + 1$, and since true for $n = 3$, by induction, true for all $n \in \mathbb{N}$ with $n > 2$.

7 $n = 1$ $f(x) = e^{px}$

LHS: $f'(x) = pe^{px}$

RHS: $f'(x) = pe^{px}$

$\therefore$ true for $n = 1$

Assume true for some $n = k$.

Then $f^{(k)}(x) = p^k e^{px}$

Consider $n = k + 1$.

$f^{(k+1)}(x) = \frac{d}{dx}(f^{(k)}(x)) = \frac{d}{dx}(p^k e^{px})$

$= p \times p^k e^{px}$

$= p^{k+1}e^{px}$

$\therefore$ If true for $n = k$ then true for $n = k + 1$, and since true for $n = 1$, by induction true for all $n \in \mathbb{N}$.

contd

8 For $n = 1$ $A^1 = \begin{bmatrix} p^1 & 0 \\ 0 & q^1 \end{bmatrix}$ $\therefore$ true for $n = 1$

Assume true for some $n = k$.

$A^k = \begin{bmatrix} p^k & 0 \\ 0 & q^k \end{bmatrix}$

Consider $n = k + 1$.

$A^{k+1} = AA^k$

$= A \begin{bmatrix} p^k & 0 \\ 0 & q^k \end{bmatrix}$

$= \begin{bmatrix} p & 0 \\ 0 & q \end{bmatrix}\begin{bmatrix} p^k & 0 \\ 0 & q^k \end{bmatrix} = \begin{bmatrix} p \times p^k & 0 \\ 0 & q \times q^k \end{bmatrix}$

$= \begin{bmatrix} p^{k+1} & 0 \\ 0 & q^{k+1} \end{bmatrix}$

$\therefore$ If true for $n = k$ then true for $n = k + 1$, and since true for $n = 1$, by induction true for all $n \in \mathbb{N}$.

9 $n = 1$ $F^{(1)}(x) = \int f(x)dx$

$\qquad\qquad = \int (1 + 2x)^1 dx$

$F^{(n)}(x) = \dfrac{(1 + 2x)^{n+1}}{(n + 1)! \, 2^n}$

When $n = 1$:

LHS $= \dfrac{(1 + 2x)^2}{2 \times 2} = \dfrac{(1 + 2x)^2}{4}$

RHS $= \dfrac{(1 + 2x)^{1+1}}{2! 2^1} = \dfrac{(1 + 2x)^2}{4}$

LHS = RHS $\therefore$ true for $n = 1$

Assume true for some $n = k$

i.e. $F^{(k)}(x) = \dfrac{(1 + 2x)^{k+1}}{(k + 1)! \, 2^k}$

Consider $n = k + 1$.

$F^{(k+1)}(x) = \int \dfrac{(1 + 2x)^{k+1}}{(k + 1)! \, 2^k} dx$

$\qquad\qquad = \dfrac{(1 + 2x)^{(k+1)+1}}{[(k + 1) + 1](k + 1)! \, 2 \times 2^k}$

$\therefore F^{(k+1)} = \dfrac{(1 + 2x)^{(k+1)+1}}{[(k + 1)+ 1]! \, 2^{k+1}}$

So if true for $n = k$ then true for $n = k + 1$.

And since true for $n = 1$, by induction true for all positive integers n.

Euclidean algorithm – pp. 98–99

1 $29\,400 = 4 \times 6860 + 1960$

$\qquad 6860 = 3 \times 1960 + 980$

$\qquad 1960 = 2 \times 980 + 0$

Hence the gcd of $29\,400$ and 6860 is 980.

$\qquad 980 = 6860 - 3 \times 1960$

$\qquad\quad = 6860 - 3(29\,400 - 4 \times 6860)$

$\qquad\quad = 13 \times 6860 - 3 \times 29\,400$

i.e. $a = -3$, $b = 13$

2 $426 = 4 \times 7^2 + 2 \times 7 + 6 \times 7^0$

$\qquad = 196 + 14 + 6$

$\qquad = 216_{10}$

$216 = 1 \times 125 + 91$

$\ 91 = 3 \times 25 + 16$

$\ 16 = 3 \times 5 + 1$

$\quad 1 = 1 \times 1$

Hence $426_7 = 216_{10} = 1331_5$

or

$\dfrac{216}{5} = 43 + \dfrac{1}{5}$, $\dfrac{216}{5^2} = 8 + \dfrac{3}{5} + \dfrac{1}{5^2}$, $\dfrac{216}{5^3} = 1 + \dfrac{3}{5} + \dfrac{3}{5^2} + \dfrac{1}{5^3}$.

Hence $426_7 = 216_{10} = 5^3 + 3 \times 5^2 + 3 \times 5 + 1 = 1331_5$.

3 $231 = 13 \times 17 + 10$

$\ 17 = 1 \times 10 + 7$

$\ 10 = 1 \times 7 + 3$

$\quad 7 = 2 \times 3 + 1$.

Hence $(231, 17) = 1$.

$1 = 7 - 2 \times 3$

$\ = 7 - 2(10 - 7) = 3 \times 7 - 2 \times 10$

$\ = 3 \times (17 - 10) - 2 \times 10 = 3 \times 17 - 5 \times 10$

$\ = 3 \times 17 - 5(231 - 13 \times 17)$

$\ = 68 \times 17 - 5 \times 231$

So, $x = -5$ and $y = 68$.